T0326614

Dynamical Properties in Nanostructured and Low-Dimensional Materials

Dynamical Properties in Nanostructured and Low-Dimensional Materials

Michael G Cottam

Department of Physics and Astronomy, University of Western Ontario
London, Ontario N6A 3K7, Canada

IOP Publishing, Bristol, UK

ISBN   978-0-7503-1054-3 (ebook)
ISBN   978-0-7503-1055-0 (print)
ISBN   978-0-7503-1112-0 (mobi)

DOI   10.1088/978-0-7503-1054-3

Version: 20151101

IOP Expanding Physics
ISSN 2053-2563 (online)
ISSN 2054-7315 (print)

British Library Cataloguing-in-Publication Data: A catalogue record for this book is available from the British Library.

Published by IOP Publishing, wholly owned by The Institute of Physics, London

IOP Publishing, Temple Circus, Temple Way, Bristol, BS1 6HG, UK

US Office: IOP Publishing, Inc., 190 North Independence Mall West, Suite 601, Philadelphia, PA 19106, USA

To my wife Sandra, for her love and also for her patience during the writing of this book.

Contents

Preface

The last few years have seen dramatic advances in the growth, fabrication and characterization of low-dimensional materials, such as graphene on the one hand, and a wide variety of nanostructures on the other, such as those formed from ultrathin films, wires, disks and other 'dots'. The elements of the nanostructures may be formed either singly or in spatial arrays that can be periodic or otherwise (in one dimension or more) and may involve more than one type of material. Most studies of these artificially engineered materials have been driven by their potential for device applications that involve smaller and smaller physical dimensions and novel utilization of the nanostructuring. In particular, the dynamical properties of these materials are of fundamental interest for devices that involve high-frequency operation and/or switching.

Consequently the different waves or excitations (vibrational, magnetic, optical, electronic and so on) need to be understood from the perspective of how their properties are modified in finite structures (especially on the nanometre to micrometre length scales) due to the presence of surfaces and interfaces. Recently the patterning of nanoelements, into periodic and other arrays, has become a focus of intense activity, leading for example to photonic crystals for optical signals (and their analogues such as phononic and magnonic crystals in other contexts). Here the control of the band gaps in the excitation spectrum is a basis for a variety of applications. Furthermore, the nonlinear properties of the excitations are increasingly a topic of interest, as well as their linear dynamics, and it is often the case that nonlinear excitations may arise that have no counterpart in the linear regime. Novel materials such as single-sheet graphene and related structures have provided the accurate realization of two-dimensional structures.

The target readership of this book includes physicists, chemists, materials scientists and engineers. They may be based in universities, industry and research laboratories. The level of presentation of this book is appropriate for final-year undergraduates, graduate students and researchers. The treatment given includes both the experimental and theoretical aspects of this expanding research field, but with an emphasis on the fundamental physical principles. Readers are assumed to have a basic background knowledge of quantum mechanics, statistical physics, electromagnetism and solid-state physics. More advanced topics are introduced and explained where needed.

The author is a Professor of Physics in the Department of Physics and Astronomy at the University of Western Ontario, Canada. He is a former Chair of the department and has also served as Associate Dean of Science (Research) and as the Director of the Western Institute for Nanomaterials Science. His research expertise is in the quantum theory of condensed-matter systems and, in particular, in the dynamical properties of the excitations, such as vibrational waves (or phonons) and magnetic excitations (spin waves or magnons), in these materials. Areas of focus in his recent work have included low-dimensional structures and nanostructured materials. He is author or co-author of around three hundred research papers and several books, including Cottam and Tilley 2004 *Introduction to Surface and Superlattice Excitations* 2nd edn (Bristol: IOP). The decade or more since that book was written has seen dramatic advances and new directions in this field, which are covered in the present publication. Notable examples are novel materials such as graphene, the development of lateral arrays of interacting elements (giving rise to emerging topics such as phononic and magnonic crystals), interest in multiferroic materials and further advances in nonlinear excitations.

I am indebted to the many friends, colleagues and collaborators who directly or indirectly have influenced this book and provided ideas.

Michael G Cottam
London, Ontario, Canada
November 2015

Abbreviations

3D, 2D, 1D, 0D	three-dimensional, two-dimensional, one-dimensional, zero-dimensional
2DEG	two-dimensional electron gas
ABC	additional boundary condition
AFMR	antiferromagnetic resonance
ARPES	angle-resolved photoemission spectroscopy
ATR	attenuated total reflection
bcc	body-centred cubic
bct	body-centred tetragonal
BEC	Bose–Einstein condensation
BG	Bleustein–Gulyaev
BLS	Brillouin light scattering
DE	Damon–Eshbach
DM	Dzyaloshinskii–Moriya
EELS	electron energy loss spectroscopy
fcc	face-centred cubic
FMR	ferromagnetic resonance
GMR	giant magnetoresistance
GNR	graphene nanoribbon
H.c.	Hermitian conjugate
HREELS	high-resolution electron energy loss spectroscopy
HP	Holstein–Primakoff (transformation)
KDP	potassium dihydrogen phosphate
LA	longitudinal acoustic
LRSP	long-range surface plasmon
MC	magnonic crystal
ML	monolayer
MQW	multi-quantum-well
n–i–p–i	(n-type)–(insulator)–(p-type)–(insulator)
NL	nonlinear
NLSE	nonlinear Schrödinger equation
OOMMF	Object Oriented MicroMagnetic Framework
PBG	photonic band gap
RKKY	Ruderman–Kittel–Kasuya–Yosida (interaction)
RPA	random-phase approximation
RS	Raman scattering
sc	simple cubic
sG	sine-Gordon
SHG	second-harmonic generation
SVA	slowly varying amplitude (approximation)
SW	spin wave
SWR	spin-wave resonance
TA	transverse acoustic
TDM	tridiagonal matrix
YIG	yttrium iron garnet

Acknowledgements

It is a pleasure to acknowledge many colleagues and research collaborators who have provided the inspiration and background for the ideas presented in this book. These include, among others, Rodney Loudon, David Tilley, Mohamed Babiker, Robert Camley, Eudenilsen Albuquerque, Peter Grünberg, Alexei Maradudin, David Lockwood, Meng Hau Kuok, Andrei Slavin, Gianluca Gubbiotti, and Giovanni Fanchini. Among my former graduate students and postdoctoral fellows, I would particularly like to mention and to thank Robert Moul, Dimosthenis Kontos, Nicos Constantinou, Niu-Niu Chen, Raimundo Costa Filho, J. Milton Pereira, Igor Rojdestvenski, Oleg Vassiliev, Eric Meloche, Trinh Nguyen, Hoa Nguyen, Tushar Das, and Arash Akbari-Sharbaf. I am grateful for the guidance provided by my editors at IOPP, initially John Navas and later Charlotte Ashbrooke, and by other editorial and production staff at IOPP.

Author biography

Michael G Cottam

Michael Cottam has received his education in the UK, obtaining his Bachelor's and Master's degrees in mathematics and physics from Cambridge University and his doctorate (DPhil) in theoretical physics from Oxford University. He worked as a research scientist in semiconductor physics for the Plessey Company (UK) before becoming a faculty member at the University of Essex. In 1987 he moved to the University of Western Ontario, Canada, where he is a Professor in the Department of Physics & Astronomy. He is a former Chair of the Department and also held appointments as Associate Dean of Science (Research) and Director of the Western Institute for Nanomaterials Science. He was recently honoured by the International Astronomical Union with the renaming of a main-belt asteroid (formerly 273262) as asteroid Cottam, in recognition of his scientific work on condensed-matter systems. His research interests over the years have spanned many aspects of the linear and nonlinear dynamics of materials, particularly on the nanoscale. He is married to Sandra Fox and they have four children/stepchildren.

IOP Publishing

Dynamical Properties in Nanostructured and Low-Dimensional Materials

Michael G Cottam

Chapter 1

Introduction

In this chapter we will introduce many of the concepts that are central to the subject matter of this book. The topics will include a preliminary survey of the different waves or excitations in solids, along with a brief discussion of their linear and nonlinear (NL) dynamical properties, some examples of the low-dimensional structures and arrays that will be of importance for later applications and also the experimental and theoretical techniques that will be needed as tools in this study.

We will be less concerned here with the *static*, or time-independent, behaviour of solids and instead focus on the temporal (and usual spatial) evolution of some characteristic property that can propagate through a solid or be localized in certain regions. Corresponding to the latter case the *excitations* mentioned above can be considered as just the wave-like Fourier components of a general propagating signal in the solid. Physically the signal might, for example, consist of optical waves, vibrations of the atoms in the solid, the fluctuating magnetization in a magnetic material, or the collective motion of charges in a plasma. In a finite-sized solid the excitations will be required to satisfy boundary conditions at the surfaces or interfaces and so their properties may become modified. The novel properties of the excitations in certain finite structures, sometimes taken in conjunction with their NL dynamics as explained later, are of fundamental interest scientifically and may lead to a wide variety of practical device applications.

1.1 Types of excitations or waves

1.1.1 Fundamentals of excitations

We start by briefly describing the 'bulk' (or 'volume') excitations that are characteristic of effectively unbounded solids, where there are no surface effects to be taken into account. Details may be found in standard textbooks on solid-state physics (see, e.g., Kittel [1], Ashcroft and Mermin [2], and Grosso and Parravicini [3]). The modifying effects of surfaces and interfaces will be introduced next by following, for example, an approach based on symmetry arguments adopted for geometries of thin films and multilayers as in the book by Cottam and Tilley [4]. From a consideration of the symmetry and crystal lattices we shall arrive, via reciprocal lattices and Brillouin zones, at a form of Bloch's theorem for the elementary excitations.

A bulk crystalline solid may be regarded as a three-dimensional (3D) repetition in space of a *basis* or building block consisting of one or more atoms. The basis can be associated with the points on an infinitely extended lattice, which may be defined from symmetry in terms of three fundamental non-coplanar translational vectors, denoted by $\mathbf{a}_1$, $\mathbf{a}_2$ and $\mathbf{a}_3$. It then follows that the vector $\mathbf{R}$ connecting any two lattice points can be expressed as

$$\mathbf{R} = n_1\mathbf{a}_1 + n_2\mathbf{a}_2 + n_3\mathbf{a}_3, \tag{1.1}$$

where n_1, n_2 and n_3 take any integer values. Of course, the lattice may have other symmetry operations, such as rotations and reflections, that also leave the space lattice unchanged. We will see that the translations are particularly important and they imply that any physical quantity $f(\mathbf{r})$ at a general position $\mathbf{r}$ in the system satisfies

$$f(\mathbf{r}) = f(\mathbf{r} + \mathbf{R}) \tag{1.2}$$

for any $\mathbf{R}$ given by equation (1.1). Depending on the context, the quantity $f(\mathbf{r})$ might refer to a component of the magnetization vector or a charge density, for example.

The *unit cell* of a lattice is just the smallest volume with the property that all space can be filled by periodic repetitions of the unit cell for each lattice. Its specification is not unique, but one simple choice is the parallelepiped formed by the basic vectors $\mathbf{a}_1$, $\mathbf{a}_2$ and $\mathbf{a}_3$. Another construction is the *Wigner–Seitz* unit cell formed by drawing the perpendicular bisecting planes of the lattice vectors from any chosen lattice point to its neighbouring sites. Details are given in, e.g., [1–3]. Some choices of unit cell are shown in figure 1.1 for the special case of a two-dimensional (2D) oblique lattice with the basic vectors $\mathbf{a}_1$ and $\mathbf{a}_2$.

The next step is to introduce the *reciprocal lattice*. The basic vectors $\mathbf{b}_1$, $\mathbf{b}_2$ and $\mathbf{b}_3$ of this lattice in reciprocal space have the properties in 3D that

$$\mathbf{a}_i \cdot \mathbf{b}_j = 2\pi\delta_{ij}, \tag{1.3}$$

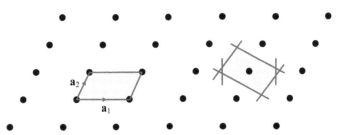

Figure 1.1. An oblique 2D space lattice with the basic vectors $\mathbf{a}_1$ and $\mathbf{a}_2$, showing two choices for the unit cell. The one on the left is a parallelepiped and the Wigner–Seitz cell is illustrated on the right.

where i and j can take values 1, 2 or 3, and δ_{ij} is the Kronecker delta (equal to unity if the subscripts have the same value and zero otherwise). For example, we may write $\mathbf{b}_1 = (2\pi/V_0)(\mathbf{a}_2 \times \mathbf{a}_3)$ and so forth, with $V_0 = |\mathbf{a}_1 \cdot (\mathbf{a}_2 \times \mathbf{a}_3)|$ denoting the volume of the unit cell for the space lattice (see, e.g., [1]). Since $f(\mathbf{r})$ introduced in equation (1.2) is a periodic function (generally in 3D) it can be expressed in terms of a Fourier series using complex exponentials as

$$f(\mathbf{r}) = \sum_{\mathbf{Q}} F(\mathbf{Q}) \exp(i\mathbf{Q} \cdot \mathbf{r}), \tag{1.4}$$

where the $\mathbf{Q}$ vectors are required to satisfy $\exp(i\mathbf{Q} \cdot \mathbf{R}) = 1$. It is easily shown [1] that they are expressible in terms of the basic vectors of the reciprocal lattice as

$$\mathbf{Q} = \nu_1 \mathbf{b}_1 + \nu_2 \mathbf{b}_2 + \nu_3 \mathbf{b}_3, \tag{1.5}$$

with ν_1, ν_2 and ν_3 denoting integers. Finally we note that the unit cell of the reciprocal lattice is known as the *Brillouin zone*. The results are particularly straightforward in the case of a simple-cubic (sc) space lattice with lattice parameter a, because the reciprocal lattice is also sc. We may choose for the space lattice

$$\mathbf{a}_1 = a(1, 0, 0), \qquad \mathbf{a}_2 = a(0, 1, 0), \qquad \mathbf{a}_3 = a(0, 0, 1), \tag{1.6}$$

whereupon the basic vectors of the reciprocal lattice are simply

$$\mathbf{b}_1 = \frac{2\pi}{a}(1, 0, 0), \qquad \mathbf{b}_2 = \frac{2\pi}{a}(0, 1, 0), \qquad \mathbf{b}_3 = \frac{2\pi}{a}(0, 0, 1). \tag{1.7}$$

Now we discuss how the translational symmetries of the space and reciprocal lattices influence the elementary excitations in an unbounded solid. The most fundamental result is embodied in *Bloch's theorem* [5], which provides information about the spatial behaviour of the variable that describes the amplitude of the excitations. This variable, which we will denote generically as a function of position as $\psi(\mathbf{r})$, depends on the type of excitation. For example, it could be the quantum-mechanical wave function in the case of electronic states, an atomic displacement in the case of lattice vibrations (or phonons), or the fluctuating magnetization in the case of spin waves (SWs) (or magnons). These and other specific examples will be discussed in more depth in later chapters after we consider some general properties here. Bloch's theorem can be expressed generally as

$$\psi(\mathbf{r}) = \exp(i\mathbf{q} \cdot \mathbf{r}) U_{\mathbf{q}}(\mathbf{r}). \tag{1.8}$$

Here $U_{\mathbf{q}}(\mathbf{r})$ is a periodic amplitude function, just as in equation (1.2), and the phase factor $\exp(i\mathbf{q} \cdot \mathbf{r})$ describes a plane-wave variation of the amplitude $\psi(\mathbf{r})$. We shall not provide a proof of this result, since this can be found in the solid-state physics textbooks cited earlier. The wave vector $\mathbf{q}$ is evidently not unique, since it can be easily checked that values differing by any reciprocal lattice $\mathbf{Q}$ as defined in equation (1.5) will equally satisfy the requirements of Bloch's theorem. Typically, as we will see later, an energy eigenvalue E will be calculated for the dynamic variable $\psi(\mathbf{r})$ using its equation of motion, e.g., in the case of an electronic excitation $\psi(\mathbf{r})$ could be the spatial wave function which satisfies Schrödinger's equation. On writing $E = \hbar\omega(\mathbf{q})$ with $\omega(\mathbf{q})$ representing the angular frequency of the excitation, it follows that we must have

$$\omega(\mathbf{q} + \mathbf{Q}) = \omega(\mathbf{q}). \tag{1.9}$$

In the idealized case of an unbounded crystal lattice the components of the vector $\mathbf{q}$ can take any real values, but equation (1.9) tells us that it is sufficient to take only $\mathbf{q}$ values that lie within the first Brillouin zone (the Wigner–Seitz cell centred at $\mathbf{q} = 0$). For the example of an sc space lattice mentioned earlier this gives $\mathbf{q} = (q_x, q_y, q_z)$ with

$$-\pi/a < q_x \leqslant \pi/a, \qquad -\pi/a < q_z \leqslant \pi/a, \qquad -\pi/a < q_z \leqslant \pi/a. \tag{1.10}$$

There are analogous results to equations (1.6), (1.7) and (1.10) for other 3D space lattices [1], such as body-centred cubic (bcc) and face-centred cubic (fcc) lattices.

While there are many 3D space lattices, there are relatively few in 2D. In fact, there are only five instead of fourteen. As we will see later, these 2D lattices will be important for thin films where there is translational symmetry in the 2D planes of atoms parallel to the surfaces, but not in the other dimension perpendicular to the surfaces. Consequently Bloch's theorem will be modified, as we discuss later. A 2D example, which is currently of considerable interest, is provided by *graphene*, which can be produced in the form of single sheets of atoms arranged in a 2D hexagonal or honeycomb space lattice (see, e.g., [6, 7]). The lattice structure is illustrated in figure 1.2 and it can be described in terms of two vectors $\mathbf{a}_1$ and $\mathbf{a}_2$ that form the basis of a unit cell with two carbon atoms per cell. There are two sublattices of carbon atoms, labelled A and B, as shown. Relative to the x- and y-coordinate axes we may write

$$\mathbf{a}_1 = \frac{\sqrt{3}\,a_0}{2}\left(\sqrt{3}, -1\right), \qquad \mathbf{a}_2 = \frac{\sqrt{3}\,a_0}{2}\left(\sqrt{3}, 1\right), \tag{1.11}$$

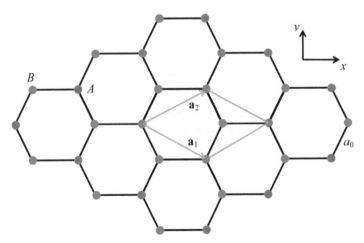

Figure 1.2. The spatial structure of a single sheet of graphene. Each carbon atom lies on one of two interpenetrating sublattices labelled A and B (shown here as red and blue circles, respectively). A possible choice for the two basic lattice vectors $\mathbf{a}_1$ and $\mathbf{a}_2$ is indicated, together with the corresponding unit cell (shown shaded).

where a_0 denotes the nearest-neighbour carbon–carbon distance. Also the basic vectors of the reciprocal lattice are simply [7]

$$\mathbf{b}_1 = \frac{2\pi}{3a_0}\left(1, -\sqrt{3}\right), \qquad \mathbf{b}_2 = \frac{2\pi}{3a_0}\left(1, \sqrt{3}\right). \tag{1.12}$$

In some other nanosystems to be considered later we might have just one direction in which there is a translational symmetry. For example, long nanowires and nanotubes that have a uniform, finite cross-sectional shape will exhibit translational symmetry with respect to their length coordinate only. Also one-dimensional (1D) linear-chain materials (such as the ferromagnetic chain pyroxene $NaCrGe_2O_6$ [8]) will have this property. By contrast, nanostructures that are finite in all three spatial dimensions will display no translational symmetry and are sometimes referred to as being zero-dimensional (0D).

1.1.2 Linear and nonlinear dynamics for excitations

Before looking further into the influence of the geometric structure on the excitations, it is useful to briefly consider how the excitations relate to the linear and NL aspects of the system dynamics. Although real physical systems might exhibit linear dynamics to a good approximation in many circumstances, it is nevertheless the case that NL behaviour will occur to some extent and may be enhanced under particular circumstances. It is to be expected that the excitations associated with the linear regime will consequently have their properties modified when nonlinearities are taken into account. It is important to enquire in what ways and to what degree this takes place and if there are practical effects to be utilized. Apart from this, we will find that there are some excitations associated with the NL regime of behaviour that have no counterpart in the linear approximation.

At this preliminary stage we will highlight some distinctions between the linear and NL dynamics of excitations, mainly by taking a few specific examples. These and other cases will then be explored in much more detail later, especially in chapter 7. Some general references covering NL dynamical phenomena, mostly for bulk properties, are the books by Yariv [9], Butcher and Cotter [10] and Mills [11] for NL optical materials and by Gurevich and Melkov [12] and Stancil and Prabakhar [13] for NL magnetic materials.

We start by considering an example of optical (electromagnetic) waves in a dielectric material, ignoring for the moment any specifically magnetic effects. The material is assumed to be homogeneous and sufficiently large that we need not explicitly take account of boundary effects. The response of the material to an external position- and time-dependent electric field $\mathbf{E}(\mathbf{r}, t)$ is conventionally described in terms of a dielectric function ε by writing

$$\mathbf{D}(\mathbf{r}, t) = \varepsilon_0 \int\!\!\int \varepsilon(\mathbf{r} - \mathbf{r}', t - t')\mathbf{E}(\mathbf{r}', t')\mathrm{d}^3\mathbf{r}'\mathrm{d}t'. \tag{1.13}$$

Here ε_0 is the permittivity of free space and $\mathbf{D}(\mathbf{r}, t)$ is the electric displacement (see, e.g., [14, 15]). The $\mathbf{D}$ and $\mathbf{E}$ vectors are also related to the polarization $\mathbf{P}$ in the material by

$$\mathbf{D}(\mathbf{r}, t) = \varepsilon_0\mathbf{E}(\mathbf{r}, t) + \mathbf{P}(\mathbf{r}, t), \tag{1.14}$$

where $\mathbf{P}(\mathbf{r}, t)$ is the polarization induced in the material. Rather than considering these quantities in terms of position and time, it is more useful for the dynamics to make Fourier transforms to a wave vector $\mathbf{q}$ and angular frequency ω according to, for example,

$$\mathbf{E}(\mathbf{r}, t) = \int\int \mathbf{E}(\mathbf{q}, \omega)\exp(\mathrm{i}\mathbf{q} \cdot \mathbf{r} - \mathrm{i}\omega t)\mathrm{d}^3\mathbf{q} \; \mathrm{d}\omega. \tag{1.15}$$

Then equation (1.13) can be re-expressed more compactly and conveniently as

$$\mathbf{D}(\mathbf{q}, \omega) = \varepsilon_0\varepsilon(\mathbf{q}, \omega)\mathbf{E}(\mathbf{q}, \omega), \tag{1.16}$$

while the corresponding result for the polarization is

$$\mathbf{P}(\mathbf{q}, \omega) = \varepsilon_0[\varepsilon(\mathbf{q}, \omega) - 1]\mathbf{E}(\mathbf{q}, \omega). \tag{1.17}$$

The $\mathbf{q}$-dependence of the dielectric function in the above two equations is known as *spatial dispersion* and arises from the fact that the dielectric term in equation (1.13) depends on the spatial separation $\mathbf{r} - \mathbf{r}'$ rather than on $\mathbf{r}$ and $\mathbf{r}'$ separately. However, in many cases this spatial dispersion may vanish or be negligible and we have simply a frequency-dependent dielectric function $\varepsilon(\omega)$ that is characteristic of the dynamical processes in the material being considered. Examples will be given later.

Before going any further we need to elaborate on two points regarding the above discussion. The first is that, in general, the vectors $\mathbf{D}$ and $\mathbf{P}$ may not necessarily be collinear with $\mathbf{E}$. The above relationships are still applicable provided the scalar dielectric function, $\varepsilon(\mathbf{q}, \omega)$ or $\varepsilon(\omega)$ as appropriate, is replaced by a second-rank tensor or equivalently a matrix in these cases. The second and more significant point is that for small values of the electric field there is a *linear* relationship, as expressed by equation (1.16), between $\mathbf{D}$ and $\mathbf{E}$. In other words, $\varepsilon(\mathbf{q}, \omega)$ is a constant function that does not depend on $\mathbf{E}$ and it is called the *linear dielectric function*. At larger electric-field values there is usually a *NL regime* in which the dielectric function becomes dependent on $\mathbf{E}$. A conventional approach in NL optics is to expand the dielectric function as a power series in terms of components of the electric-field components, as we describe in a later chapter. Various different NL effects can then be identified depending on which terms in the expansion are important. For example, the *Kerr-type nonlinearity* involves an effective NL dielectric function [10, 11] of the form

$$\varepsilon^{\mathrm{NL}} = \varepsilon + \rho \, |\mathbf{E}|^2. \tag{1.18}$$

Here both the linear part ϵ and the coefficient ρ before the field-dependent term may, in general, be functions of wave vector $\mathbf{q}$ and frequency ω.

To probe further some distinctions between the linear and NL dynamics arising due to the form of the dielectric function, we now turn to the basic equation satisfied by the electric field. This is just the electromagnetic wave equation arising from Maxwell's equations [14, 15]:

$$\varepsilon(\omega)\frac{\partial^2\mathbf{E}}{\partial t^2} - c^2\nabla^2\mathbf{E} = 0. \tag{1.19}$$

This result is the form appropriate for a transverse wave (i.e., when the wave vector $\mathbf{q}$ specifying the propagation direction is perpendicular to $\mathbf{E}$) and c is the velocity of light in a vacuum. We note that the solution of this equation in the linear regime is relatively straightforward, because the linear dielectric function does not depend on $\mathbf{E}$ and so equation (1.19) is a linear differential equation. We are ignoring spatial dispersion here. The solution in this case where we assume an unbounded medium can be found in a plane-wave $\exp[i(\mathbf{q} \cdot \mathbf{r} - \omega t)]$ form, yielding

$$q^2 = \varepsilon(\omega)\omega^2/c^2. \tag{1.20}$$

This is just the implicit dispersion relation (ω versus q relationship) arising when the electromagnetic wave is coupled to the polarization field of the medium. The excitation is generally referred to as a *polariton*. If the dielectric function is just a constant and written as n^2, where n is the refractive index of the medium, we recover the usual dispersion relation $\omega = (c/n)q$ for a photon (light) with speed c/n in the medium. By contrast, a simple example of a frequency-dependent dielectric function occurs for an electron plasma, as for the gas of electrons in a metal or semiconductor in the presence of positive ions that ensure the overall charge neutrality. The plasma oscillations [1, 2] of the electron gas are known to define a characteristic angular frequency, the plasma frequency ω_p, given by $\omega_p^2 = n_0 e^2/\varepsilon_0 m$ where n_0 is the 3D concentration of electrons, each having mass m and charge $-e$. It can be shown (see also chapter 4) that

$$\varepsilon(\omega) = \varepsilon_\infty\left(1 - \frac{\omega_p^2}{\omega^2}\right), \tag{1.21}$$

where ε_∞ is the limiting dielectric constant at frequencies well above ω_p and damping has been neglected. Substituting this into the expression in equation (1.20) and rearranging gives the plasmon-polariton dispersion relation explicitly as

$$\omega = \sqrt{\omega_p^2 + c^2 q^2}. \tag{1.22}$$

This simple dispersion relation is drawn as the solid line in figure 1.3. We have also included for comparison the dashed line $\omega = cq$ which corresponds to the light line for propagation in a vacuum. There are two limiting regions of interest for the polariton. One is for $cq/\omega_p \ll 1$ when $\omega \simeq \omega_p$ and the other is for $cq/\omega_p \gg 1$ when $\omega \simeq cq$. There is also a band gap of forbidden frequencies where no plasmon-polaritons exist and this occurs here in the frequency range from 0 to ω_p.

The above discussion relates to an example of a *bulk* (or *volume*) excitation, since no account has been taken of any surface or boundary and the linear excitation propagates (or is wave-like) in all three spatial dimensions. Other linear types of polaritons may occur that are associated with surfaces and interfaces, and this behaviour will be one of the main topics in chapter 5.

We discussed earlier how NL effects could arise, for example, through a Kerr-type form of the dielectric function, as in equation (1.18). Before concluding this section we will outline a case of a NL polariton due to this term; more details will be given in chapter 7 (but see also [4]). This example is related to the phenomenon

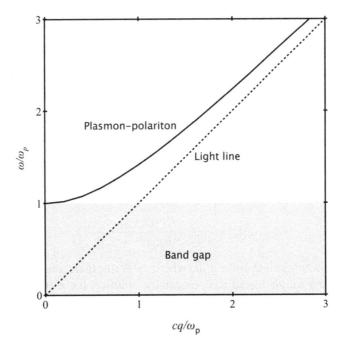

Figure 1.3. Dispersion relation for a linear plasmon-polariton (solid line) in a 3D material, as given by equation (1.22). The angular frequency, in terms of dimensionless ω/ω_p is plotted versus dimensionless wavenumber, in terms of cq/ω_p.

known as *self-guiding* and can occur for polaritons at an interface (see [16, 17] for reviews). We start with a qualitative description by recalling that within linear optics the guiding condition for a light wave to remain localized (through multiple internal reflections) within a film is that the refractive index of the film should exceed the refractive index of the bounding medium on each side [14, 15]. It might therefore be expected that, provided (i) ε and ρ in equation (1.18) are both positive and (ii) the electric-field term $|E|^2$ can be localized with a large value near a surface, the situation will be broadly analogous to a guiding film with a large refractive index. The above conditions can, in fact, be realized as we now demonstrate by considering an interface between a linear medium and a Kerr NL medium.

Specifically we will assume that the NL medium fills the lower half-space $z < 0$, while a linear medium occupies the upper half-space $z > 0$. The interface between them corresponds to the plane $z = 0$ and the system is assumed to be infinite in the x- and y-coordinate directions. We look for solutions for which the electric field is a s-polarized wave, which means that it can be written as $\mathbf{E} = (0, E, 0)$ with its propagation being along the x-direction. From Bloch's theorem and assuming an angular frequency ω it follows that all electromagnetic-field quantities are proportional to $\exp(iq_x x - i\omega t)$. Substitution of equation (1.18) into the wave equation (1.19) yields

$$\frac{\mathrm{d}^2 E}{\mathrm{d}z^2} - \left(\kappa^2 - \rho\frac{\omega^2}{c^2}E^2\right)E = 0, \tag{1.23}$$

where we define κ, which is real for large enough q_x, by

$$\kappa^2 = q_x^2 - \varepsilon\omega^2/c^2 \tag{1.24}$$

and we assume solutions exist where E is real. The NL differential equation can then be solved by multiplying each side of (1.23) by dE/dz and integrating to obtain

$$\left(\frac{dE}{dz}\right)^2 - \kappa^2 E^2 + \rho\frac{\omega^2}{2c^2}E^4 = 0. \tag{1.25}$$

Here a constant of integration has been set to zero on the assumption that both E and dE/dz tend to zero as $z \to -\infty$ inside the NL medium. Next, equation (1.25) can be rearranged as an expression for dE/dz as a function of z and directly integrated, giving

$$E = \frac{2^{1/2}c\kappa}{\rho^{1/2}\omega\cosh[\kappa(z + z_0)]} \tag{1.26}$$

as the solution appropriate for $z < 0$, where z_0 is another constant of integration. For completeness we also need to consider the solution for E in the linear medium (labelled 1) for $z > 0$. By contrast with equation (1.23) we now have a much simpler second-order linear differential equation, for which the solution satisfying the requirement $E \to 0$ as $z \to \infty$ is

$$E = E_1\exp(-\kappa_1 z). \tag{1.27}$$

Here $\kappa_1 = (q_x^2 - \varepsilon_1\omega^2/c^2)^{1/2}$ by analogy with equation (1.24), with ε_1 denoting the dielectric constant of medium 1 and as before we take κ_1 to be real. The constant E_1 is easily found from the electromagnetic boundary condition at the interface $z = 0$ that the tangential electric-field components must be continuous. We are still left, however, with the problem of finding the other constant z_0.

This kind of situation is typical of many problems involving NL excitations and in the present case it can be handled by finding another equation involving z_0 from consideration of the power flow in the propagating wave. In electromagnetism this comes from the time-averaged Poynting vector [14, 15] and further discussion will be deferred to chapter 7. However, we have demonstrated here by the form of equation (1.26) that a spatially localized electromagnetic wave can occur due to the non-linearity of the medium.

1.2 Survey of types of nanostructures

In the examples given so far, we have considered excitations in unbounded (bulk) media, in a 2D sheet (for graphene), or at the planar interface between two media. More generally there may be multiple boundaries (surfaces or interfaces) that affect the dimensionality and symmetries of the system, thus modifying the form of Bloch's theorem. Also, the boundaries introduce different length scales and here we will be mainly interested in effects in the range of nanometres (nm) up to micrometres (μm). We will now describe briefly some of the different nanostructures that will be considered in the following chapters. Examples are represented schematically in figure 1.4.

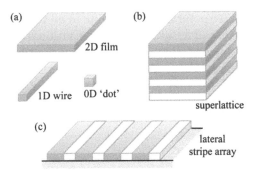

Figure 1.4. Schematic illustration of some types of nanostructures: (a) low-dimensional structures, (b) periodic multilayers or superlattices and (c) laterally patterned periodic structures.

1.2.1 Low-dimensional structures

In this category we have materials with reduced dimensionality as expressed in terms of the number of spatial dimensions in which there is translational symmetry. They include the 2D cases of thin films with finite thickness L and parallel planar surfaces, as well as sheets of atoms such as the graphene example mentioned earlier. They are characterized by a 2D space lattice associated with the planes of atoms parallel to the surfaces, as mentioned in subsection 1.1.1, and the form of Bloch's theorem will be modified slightly compared with equation (1.8) to become

$$\psi(\mathbf{r}_\parallel, z) = \exp\!\left(i\mathbf{q}_\parallel \cdot \mathbf{r}_\parallel\right) U_{\mathbf{q}_\parallel}(\mathbf{r}_\parallel, z). \tag{1.28}$$

We have assumed that the z-axis is chosen to be perpendicular to the surfaces, so the in-plane vectors are $\mathbf{r}_\parallel = (x, y)$ and $\mathbf{q}_\parallel = (q_x, q_y)$. The challenge in solving for the excitations in films is finding the z-dependence of the amplitude term $U_{\mathbf{q}_\parallel}(\mathbf{r}_\parallel, z)$.

In the case of linear excitations in films we will often find later that this z-dependence is expressible as a linear combination of terms having the form $\exp(i q_z^{(j)} z)$, where $q_z^{(j)}$ may be *complex* with a series of allowed values labelled by j (see, e.g., [4]). There are two possible situations that may arise. One is that $q_z^{(j)}$ is real and thus it represents the third wave-vector component of a bulk excitation in the film with total wave vector $\mathbf{q} = (\mathbf{q}_\parallel, q_z^{(j)})$. This bulk excitation is affected by the film surfaces because it must satisfy boundary conditions there. The other possibility is that $q_z^{(j)}$ may have an imaginary part, which we denote as $i\kappa$. Then the $\exp(i q_z^{(j)} z)$ term will involve a real factor such as $\exp(-\kappa z)$, which is characteristic of a localized surface excitation that decays with distance from one of the film surfaces. A surface excitation can therefore be typically described in terms of a 2D wave vector along with a decay length that is equal to κ^{-1}.

The category of low-dimensional structures also includes nanowires (or nano-rods), which have just one direction of translational symmetry (the axis along the length of the wire) and therefore the excitations are characterized by a 1D wave vector. The straightforward modification of Bloch's theorem in this case yields

$$\psi(x, y, z) = \exp\!\left(i q_z z\right) U_{q_z}(x, y, z), \tag{1.29}$$

where it is assumed here that the z-axis is along the translational symmetry axis.

Finally, we may have structures that are finite in all three dimensions, such as a small cuboid or a sphere. These have no translational symmetry (so Bloch's theorem does not apply) and they may be labelled as being 0D, such as the 'quantum dots' that have been of particular interest for semiconductors or flat 'platelets' or 'disks' for magnetic materials (see, e.g., [18, 19]). The low-dimensional structures mentioned here are illustrated schematically using examples in figure 1.4(a).

1.2.2 Multilayers and superlattices

If more than one material is utilized, then more elaborate cases of nanostructures can arise than those in the previous subsection. One simple possibility is to build up multilayers with alternating thin films of two materials, denoted as A and B, to form an $ABABAB\cdots$ pattern of growth on a substrate material. Such an example is provided in figure 1.4(b) for the materials shown in different colours. The simplest case would be a bilayer AB, but multilayers with a large number of layers (e.g., hundreds or more) can be fabricated. In all cases these systems would have the property of translational symmetry for lattice directions parallel to the planar surfaces and interfaces. Thus the 2D Bloch's theorem in equation (1.28) still applies, with the only essential difference being that the term $U_{q_\parallel}(r_\parallel, z)$ may have a more complicated dependence on z. There may, for example, be excitations that are localized only near the interfaces between A and B layers.

Another aspect of multilayers that makes them of special interest occurs when all the A layers have identical properties (composition, thickness, etc) to one another and likewise for all the B layers. The structure then consists of repeats of the same basic AB building block. If there is a large number of these repeats, then a new symmetry operation emerges corresponding to translations in the z-direction through a periodicity length $D = d_A + d_B$, where d_A and d_B denote the thicknesses of the individual A and B layers. We then have an example of a *periodic superlattice*, for which the inbuilt artificial periodicity in the z-direction may be associated with a new 1D Brillouin zone that extends in reciprocal space from $-\pi/D$ to π/D. Since D is typically much larger than the lattice parameter a of either constituent material, this new Brillouin zone is smaller than the crystal Brillouin zone. This can have striking consequences for the excitations that propagate as waves in the perpendicular (z-)direction of the superlattice. A smaller Brillouin zone can sometimes be advantageous as regards the experimental techniques for studying the excitations and it also leads to a 'folding' of the dispersion curves, as we will discuss through examples in section 1.5 and later chapters.

The concept of a superlattice was originally introduced by Esaki and Tsu [20] for ultrathin semiconductor multilayers where the layer thicknesses were less than the electron mean free path, but currently the term is applied more generally for multilayers that are either periodic or are generated according to certain mathematical rules. In the former category we may also have periodic arrangements with three (or more) components, $ABCABCABC\cdots$ for example, as well as the two-component superlattice described above. In the latter category we may include *quasi-periodic superlattices* (see, e.g., [21], where the sequencing of layers along the growth direction is quasi-periodic in the sense that it involves two different fundamental periods whose ratio is an irrational number. One way to achieve this is through the Fibonacci sequence of numbers, $\{F_n\}$ with integer $n = 0, 1, 2, \ldots$, that is defined by the recursion relation

$$F_{n+2} = F_{n+1} + F_n, \tag{1.30}$$

with the initial conditions $F_1 = F_0 = 1$. It is straightforward to prove that as n becomes very large ($n \to \infty$) the ratio F_{n+1}/F_n tends to the value $(1 + \sqrt{5})/2 \simeq 1.618$ which is known as the golden mean. For the application to Fibonacci superlattices some fundamental property of the layered growth is developed by analogy with equation (1.30), as explained later. The excitations in quasi-periodic superlattices have novel properties (e.g., regarding their localization, multifractal behaviour, etc) that make them of special interest [22].

1.2.3 Laterally patterned structures

By contrast with the previous subsection we will now discuss some artificial nanostructures produced by a lateral arrangement (usually periodic) of elements such as that represented in figure 1.4(c). This shows an alternating sequence of stripes of two different materials deposited side-by-side on a planar surface, rather than stacked vertically. Interest in such structures was given impetus in 1987 through seminal work by Yablonovitch [23] and John [24], who studied artificial electromagnetic materials, typically in low dimensions, that have periodic modulation of the refractive index. This was originally undertaken with a view to control the propagation, localization and scattering of light, and led to periodic structures known as either *photonic crystals* or alternatively *photonic band-gap materials* for applications in optoelectronics. Some consequences for the photonic band structure in periodic cases are developed in section 1.5.

   More recently, aided by advances in materials growth and fabrication on the sub-micron scale, other types of laterally patterned materials have been utilized to study phonons (hence *phononic crystals*), SWs or magnons (*magnonic crystals*; MCs) and plasmons (*plasmonic crystals*), among others. One of the characteristic features of these materials is the tailoring of the band structure of the excitations and the forbidden-frequency gaps (or band gaps) in the spectrum. We shall discuss these cases in more detail in later chapters. Just as with the superlattices in the previous subsection, effects will arise due to a modified Bloch's theorem and mini-Brillouin zones due to the artificial periodicity. The example depicted in figure 1.4(c) with a striped array is a 1D crystal in the lateral plane, but 2D patterning (as with a chess or checker board array) provides another case of interest.

1.3 Experimental techniques for dynamic properties

Although a wide range of different physical phenomena will contribute to the dynamic properties of materials, many of the same experimental techniques will be applicable. This will also be the case as regards the theoretical tools needed to study the excitations and so it is useful to give a broad introduction in this chapter and to provide some general references. The experimental aspects are covered first in this section and the theoretical topics will be described in section 1.4.

We will emphasize here the techniques that are most useful as regards the dynamical properties and characterization of the excitations. These include inelastic light scattering, inelastic particle scattering, electromagnetic techniques and magnetic (or spin) resonance methods, etc. A fuller account can be found in [4]. We do not describe here the sample growth and preparation techniques, or the analytical tools for surface and interface structural properties, since these are not the focus of this book and they have been extensively covered elsewhere (see, e.g., [19, 25, 26]).

Both Raman and Brillouin scattering of light by dense media involve an inelastic process whereby the incident light of angular frequency ω_I (or a photon of energy $\hbar\omega_I$) and wave vector k_I interact with the medium by creating or absorbing an excitation and producing a scattered light beam. Although these effects were discovered in the early part of the 20th century, it was not until much later with the advent of the laser and other technical developments that they were widely used for studying bulk and surface excitations. The essential difference between the two techniques lies in the method used to analyse the frequency of the scattered light; this involves a grating spectrometer for Raman scattering (RS) (where the wavenumber shifts of the light are typically in the range 5–4000 cm^{-1}) and a Fabry–Pérot interferometer for Brillouin scattering (where the wavenumber shifts are below about 5 cm^{-1}). Note that a wavenumber shift of 1 cm^{-1} corresponds to a 29.98 GHz frequency shift. We label the excitation by ω and $\mathbf{q}$ and the scattered light by ω_S and $\mathbf{k}_S$, and the two possible processes (known as Stokes and anti-Stokes scattering) are represented schematically in figure 1.5. When the light scattering takes place in a bulk (effectively infinite) 3D transparent material the energy and momentum are conserved and so

$$\hbar\omega_I = \hbar\omega_S \pm \hbar\omega, \qquad \hbar\mathbf{k}_I = \hbar\mathbf{k}_S \pm \hbar\mathbf{q}, \qquad (1.31)$$

where the upper (lower) signs correspond to the Stokes (anti-Stokes) case. The dispersion relations for the incident and scattered light yield $|k_I| = \eta_I\omega_I/c$ and $|\mathbf{k}_S| = \eta_S\omega_S/c$, where η_I and η_S denote the corresponding values of the refractive index. The above relations, together with the conservation properties in equation (1.31), imply that only excitations with wave vectors near the centre of the Brillouin zone in a bulk material are excited.

There are several modifications for the application of light scattering in low-dimensional and/or optically absorptive materials. First, with thin films and

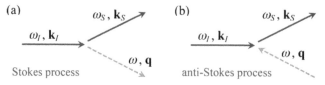

Figure 1.5. Representation of the (a) Stokes and (b) anti-Stokes processes of inelastic light scattering. The incident and scattered photons are depicted by solid lines and the created or absorbed excitation by a dashed line.

multilayers on the one hand, or nanowires and nanowire arrays on the other, the conservation of momentum will apply only in the 2D or 1D direction(s) where the system has translational symmetry. In the other dimension(s) there will be a spread of wave-vector components for the excitation, implying a spread of frequencies in the measured spectrum. The extent to which this occurs depends on the optical absorption (or imaginary part of the refractive index) in the material. If the absorption is large, which is typical for metals and some semiconductors, the light will penetrate only close to the surface of the material, so light scattering can be sensitive to surface properties and to the excitations localized there. In the case of 0D materials (or dots) none of the wave-vector components will be conserved. As well as providing information about the frequencies, light scattering also probes the intensities of the excitations. The scattering cross section is, in fact, related to the correlations between components of the electric field for the scattered light and this provides a connection between theory and experiment, as we describe later. Some general references for inelastic light scattering are [27] (mainly in bulk materials), [28, 29] (mainly in thin films and multilayers) and [30] (mainly in other low-dimensional materials).

Next we turn to the inelastic scattering of particles by the crystal excitations. In principle this process may involve larger momentum changes than encountered with light, so by contrast particle scattering may be useful for probing excitations throughout the entire Brillouin zone, depending on the type of particle involved. In electron energy loss spectroscopy (EELS), and in its high-resolution variant (HREELS), a beam of electrons is scattered from the surface of a material, as reviewed in [31, 32]. A typical geometry for the scattering of particles from a planar surface is shown in figure 1.6, where a surface excitation such as a phonon is emitted or absorbed in the process. The condition for conservation of energy yields

$$E_I = E_S \pm \hbar\omega, \qquad (1.32)$$

where we follow the notation in figure 1.6 and the upper (lower) sign refers to the creation (annihilation) case for the excitation. The other scattering condition corresponds to the conservation of the momentum component in the direction parallel to the surface plane, giving

$$|\hbar k_I \sin\theta_I - \hbar k_S \sin\theta_S| = \hbar q_\parallel. \qquad (1.33)$$

Also, we have the classical energy–momentum relations $E_I = (\hbar k_I)^2/2M$ and $E_S = (\hbar k_S)^2/2M$ with M denoting the particle mass. Generally for EELS or

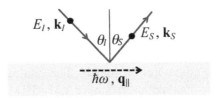

Figure 1.6. Geometry for inelastic particle scattering from a planar surface of a material, where the incident and scattered beam are in the same vertical plane.

HREELS we can approximate by assuming $E_I \gg \hbar\omega$, which is usually well satisfied (e.g., if $E_I \sim 100$ eV and $\hbar\omega \sim 100$ meV), implying $|\mathbf{k}_S| \simeq |\mathbf{k}_I| \equiv k_I$. For the same reason we can also ignore Umklapp processes in equation (1.33). It follows that, by varying the angles θ_I and θ_S, the in-plane wave vector can be varied across its 2D Brillouin zone and the ω-versus-$\mathbf{q}_\parallel$ dispersion relation for the surface excitation can be investigated. An example of this type of analysis is provided by the work of Lehwald et al [33] for phonons localized at a Ni surface. Spin-polarized EELS has been used for magnetic excitations at surfaces [34]. It involves employing a spin-polarized incident electron beam together with a polarization analyser in the scattered beam.

The inelastic scattering of particles that are much heavier than electrons allows for larger momentum changes and hence offers the prospect of making it easier (in principle) to study excitations with large wave vectors. Inelastic neutron scattering is well established for studying bulk excitations. However, neutrons interact only very weakly with materials and have macroscopically large penetration depths, so this method is relatively insensitive to surface effects. This drawback may be mitigated if the surface or interface area is effectively increased (e.g., by using a short-period superlattice) or if a neutron beam is used at grazing incidence to a surface (see, e.g., [35]). However, in the context of *surface* excitations, the inelastic scattering of small neutral atoms, such as He atoms, has proved to be more useful. The He atoms at suitable incident energies (e.g., of order 20 meV) will be scattered back from a surface by essentially just the first monolayer (ML) of the material. Also the atoms in the incident beam will have a de Broglie wavelength $(=h/\sqrt{2ME_I})$ that is typically comparable with the lattice constant, so the entire 2D Brillouin zone can be studied. The technique was pioneered by Brusdeylins et al [36] using time-of-flight spectroscopy to study surface phonons. The basic equations for the scattering are essentially the same as for EELS, equations (1.32) and (1.33), except that it is no longer true that $k_I \simeq k_S$ because of the larger particle mass here. In the case of a $90°$ scattering geometry (when $\theta_I + \theta_S = \pi/2$ in figure 1.6) it is straightforward to show that ω and $q_\parallel$ are related by

$$\omega = \frac{\hbar^2 k_I{}^2}{2M}\left[\left(\frac{q_\parallel + k_I \sin\theta_I}{k_I \cos\theta_I}\right)^2 - 1\right]. \tag{1.34}$$

This allows the dispersion relation of a surface excitation to be deduced.

To conclude this section we briefly mention some experimental techniques that are specific to certain types of excitations. Others will be mentioned in later chapters. Magnetic resonance techniques are widely used for excitations in magnetic materials, both in the bulk and in low-dimensional geometries (see [29] for a review). At low power levels a microwave magnetic field (in a direction transverse to the magnetization) will couple linearly to the zero wave-vector SWs (magnons) in a ferromagnet, leading to the ferromagnetic resonance (FMR) condition when the frequencies of the field and excitation match. The analogous effect in antiferromagnets is called antiferromagnetic resonance (AFMR) and typically occurs at higher frequencies. In SW resonance (SWR) in ferromagnetic thin films the technique is used to excite selectively the standing SWs across the film thickness that have small but nonzero wave vectors. At higher power levels the oscillating (or 'pumping') magnetic field can be used to promote various NL SW processes in ferromagnets, as we will discuss in chapter 7.

Another area involving specialized techniques occurs with excitations that generate electromagnetic waves. These can be studied using optical methods that employ, for example, evanescent wave coupling in the region near closely positioned interfaces between media of different refractive index. This is the basis of the method of attenuated total reflection (ATR), which has been successfully applied to plasmons and polaritons (see chapters 4 and 5). Evanescent wave properties are also exploited in scanning near-field optical microscopy, which is a form of scanning probe microscopy.

1.4 Theoretical methods for dynamic properties

1.4.1 Equations of motion for excitations

Some of the theoretical considerations have already been mentioned in earlier parts of this chapter, particularly in subsection 1.1.2, where we presented some simple examples of linear and NL excitations. There we emphasized the role of an equation of motion for the excitation amplitude, e.g., Maxwell's wave equation for the electric field as in equation (1.19), along with boundary conditions and/or the use of Bloch's theorem to deal with spatial periodicity where needed.

From such an approach one would normally be able to deduce the excitation frequency and hence a dispersion relation, as well as information about the spatial variation in the case of a localized excitation. As discussed earlier in section 1.2, a bulk excitation will depend on a 3D wave vector $\mathbf{q}$, whereas a surface or interface excitation at a planar interface will be characterized by a 2D wave vector $\mathbf{q}_\parallel$ and one or more attenuation factors (or decay lengths). Other considerations apply in 1D and 0D systems, as mentioned before. Calculating the excitation frequencies in a low-dimensional system normally involves having a set of *homogeneous* equations of motion satisfied by the amplitude variable(s) in the different parts of the system. These equations may either be differential equations if a macroscopic (or continuum) approach is being followed, as in the previous examples given for electromagnetic excitations, or finite-difference equations if a microscopic approach is used, as in a discrete-lattice model. Both approaches will be used throughout the following chapters. Also, in some cases, numerical simulation packages have been developed to carry out these calculations. An example is the Object Oriented MicroMagnetic Framework (OOMMF) project at the US National Institute of Standards and Technology, which is applicable to studying a range of properties of micromagnetic low-dimensional systems [37].

It is useful to briefly consider some other examples that involve a homogeneous equation of motion, apart from the electromagnetic one given previously. First we take the case of an infinite, isotropic elastic medium with a longitudinal vibrational wave (an acoustic phonon) propagating in one direction, which we label as the z-direction. If we denote $u(z, t)$ as the local displacement in the z-direction of the medium, ρ as the uniform density of the medium and v_L as the longitudinal sound wave velocity, the usual linear wave equation with weak damping (constant $\gamma > 0$) included is (see, e.g., [1, 3])

$$\rho\frac{\partial^2 u}{\partial t^2} + \rho\gamma\frac{\partial u}{\partial t} - \rho v_L{}^2\frac{\partial^2 u}{\partial z^2} = 0. \tag{1.35}$$

A solution is easily found by taking $u(z, t) \propto \exp[i(qz - \omega t)]$ which yields the condition $\omega^2 + i\omega\gamma = v_L{}^2 q^2$. In the limit of zero damping ($\gamma \to 0$) this gives the usual acoustic-phonon dispersion relation $\omega = v_L q$, as expected.

When a discrete-lattice approach is followed, leading to finite-difference equations for the spatial variables instead of differential equations, some special techniques are useful. One of these is the *tridiagonal matrix* (TDM) method, which is useful for surface and interface problems when the interactions are short range, often just between nearest neighbours. A general outline of the method is given in the appendix and some applications are discussed later, e.g., for the lattice dynamics of phonons in chapter 2 and for exchange-dominated SWs in ferromagnets in chapter 3.

A somewhat different situation will occur if the excitation is described in terms of a quantum-mechanical operator, rather than a classical amplitude as above. This will be so, for example, in the case of a magnetic system consisting of a lattice of atoms where each atom is associated with a spin angular-momentum operator (see chapter 3). If the quantum-mechanical system is represented by a Hamiltonian H, then the equation of motion for any operator A of the system is obtained from [38]

$$i\hbar\frac{dA}{dt} = [A, H], \tag{1.36}$$

where $[A, H] \equiv (AH - HA)$ denotes the commutator between the two operators. For simplicity we shall henceforth usually take units such that $\hbar = 1$. Equation (1.36) may be either a linear or NL equation in A, depending on whether the outcome of evaluating the commutator of the operator A (for the excitation) with the Hamiltonian gives an expression that is linear in A or has some other more complicated dependence. Examples of both cases will arise later in this book, but it is probably most clearly illustrated for magnons in ferromagnets (see chapters 3 and 7).

1.4.2 Linear-response theory for excitations

In many cases we are interested in other properties of the excitations apart from their dispersion relations. For example, we may need to know the relative statistical weighting (or spectral intensity) of the excitations or to find a scattering cross section to compare with experiment. This information can be provided by linear-response theory and we will now outline this method and then illustrate it with an example. The basis of the method is to calculate the response of the system to a small applied stimulus, taking care to choose the appropriate classical or quantum variables (relating to the excitation being studied). The *response functions* provide not only the dispersion relations for the relations but also the intensity-related information mentioned above. General accounts of linear-response theory are to be found, for example, in [4, 39] where the connection to Green functions is also explained.

A comprehensive treatment in the context of low-dimensional systems can be found in [40]. Here (and also in the appendix) we will present, for later reference, only the important results.

We suppose that a time-dependent perturbation described by Hamiltonian $H_1(t) = -Bf(t)$ is applied to the system, where $f(t)$ is a scalar external field that couples linearly to a system variable denoted by the operator B. An example is when B is a displacement from equilibrium in an elastic medium and $f(t)$ is a mechanical force. The total Hamiltonian can be written as

$$H = H_0 - Bf(t), \tag{1.37}$$

where H_0 denotes the Hamiltonian for the unperturbed system, taken to be in thermal equilibrium at time $t = -\infty$. At any later time the response of the system can be expressed in terms of the change $\bar{A}(t)$ produced in any other system variable corresponding to an operator denoted as A. The procedure for doing this is outlined in the appendix and the main conclusion is that

$$\bar{A}_\omega = \langle\langle A; B\rangle\rangle_\omega F(\omega), \tag{1.38}$$

where $\bar{A}_\omega$ and $F(\omega)$ are defined to be the frequency Fourier components of $\bar{A}(t)$ and $f(t')$, respectively. This result holds within a linear approximation in terms of the perturbation term and the proportionality factor written as $\langle\langle A; B\rangle\rangle_\omega$ in equation (1.38) is identified as the Fourier component of the Green function that describes correlations between the operators A and B. The above result establishes the role of Green functions as linear-response functions, in the sense that they are just the same as the proportionality factor between an applied stimulus to the system, represented by $F(\omega)$, and the response that it produces, represented by $\bar{A}_\omega$. The formal definitions and some basic properties of Green functions are given in the appendix.

In order to illustrate the formalism for response functions and Green functions, we return to the physical example of excitations in an infinite elastic medium considered earlier in this section. The homogeneous equation of motion for the displacement $u(z)$ for a longitudinal acoustic (LA) wave in the z-direction was given by equation (1.35), to which we now add a driving term (a perturbation) which is chosen to represent a harmonic point force at position z' in the system, i.e., we take

$$f(z, t) = (f_0/A_0)\exp(-i\omega t)\delta(z - z'), \tag{1.39}$$

where f_0 is an amplitude and A_0 is the area presented by the system in the xy-plane. The total interaction energy with the system is

$$H_1(t) = -\int\int\int u(z)f(z, t)dx\, dy\, dz = -f_0 \exp(-i\omega t)\int u(z)\delta(z - z')dz$$
$$= -f_0 \exp(-i\omega t)u(z'), \tag{1.40}$$

so it takes the general form assumed earlier. The equation of motion on generalizing (1.35) becomes

$$\rho\frac{\partial^2 u}{\partial t^2} + \rho\gamma\frac{\partial u}{\partial t} - \rho v_L^2\frac{\partial^2 u}{\partial z^2} = \frac{f_0}{A_0} \exp(-i\omega t)\delta(z - z'). \tag{1.41}$$

As anticipated, the outcome is that we now have an *inhomogeneous* differential equation. We may first take a common time dependence such as $\exp(-i\omega t)$, leaving as the equation for the spatial dependence

$$\frac{\mathrm{d}^2 u}{\mathrm{d}z^2} + q^2 u = -\frac{f_0}{\rho v_\mathrm{L}^2 A_0}\delta(z - z'), \tag{1.42}$$

where the complex q^2 is defined in the same way as following equation (1.35). We seek the solution that is bounded as $z - z' \to \pm\infty$ and has the correct behaviour as $z - z' \to 0$. It may easily be verified that the result is

$$u(z) = \frac{\mathrm{i}f_0}{2\pi\rho v_\mathrm{L}^2 A_0}\exp(\mathrm{i}q|z - z'|). \tag{1.43}$$

In the spirit of linear-response theory this is the driven equation for $u(z)$ due to a harmonic force applied at point $z = z'$. From equation (1.38) we obtain the Green function as

$$\langle\langle u(z); u(z')\rangle\rangle_\omega = \frac{\mathrm{i}}{2\pi\rho v_\mathrm{L}^2 A_0}\exp(\mathrm{i}q|z - z'|). \tag{1.44}$$

Since the above result is a function of $z - z'$, as expected for a system with translational symmetry, we can re-express it by making a Fourier transform from the spatial variables to a wave-vector component k in the z-direction:

$$\langle\langle u; u)\rangle\rangle_{k,\omega} \equiv \int_{-\infty}^{\infty} \langle\langle u(z); u(z')\rangle\rangle_\omega e^{-\mathrm{i}k(z-z')}\mathrm{d}(z - z') = \frac{1}{2\pi A_0\left(v_\mathrm{L}^2 k^2 - \omega^2 - \mathrm{i}\omega\gamma\right)}. \tag{1.45}$$

The last step for this result is accomplished by splitting the range of integration into two parts with $z < z'$ and $z > z'$. We note that the condition for the energy denominator to be zero yields the dispersion relation $\omega_k = v_\mathrm{L}k$ for the excitation (an acoustic phonon) if the damping is ignored. This illustrates a general property for the denominator of Green functions. Finally we may use equation (1.45) together with the fluctuation-dissipation theorem in the appendix to obtain $\langle|u|^2\rangle_{k,\omega}$ which is a measure of the spectral intensity (or mean-square amplitude) of the excitation as a function of k and ω. A numerical example of the result appears in figure 1.7, where the peaks occur for ω close to $\pm\omega_k$ provided the damping is relatively small ($\gamma \ll \omega_k$). The peaks are unequal in height due to the statistical-mechanical weighting, which depends on the temperature T. This has consequences, for example, regarding the difference in intensities observed between Stokes and anti-Stokes peaks in inelastic light scattering, as we will discuss in chapters 2 and 3.

1.5 Photonic band gaps in periodic structures

We have already briefly mentioned periodic structures in subsections 1.2.2 and 1.2.3, which are typically formed from two or more materials, either as superlattices or as lateral arrays. The bands of excitations in such structures (i.e., their frequency versus wave-vector relationships) can be very complicated in general and particular

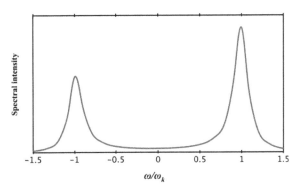

Figure 1.7. The spectral intensity $\langle |u|^2 \rangle_{k,\omega}$ versus reduced frequency ω/ω_k for the acoustic phonons in an infinite elastic medium (see the text). We denote $\omega_k \equiv v_L k$ for a fixed value of wave-vector component k. The assumed numerical values of the parameters correspond to dimensionless $k_B T/\hbar\omega_k = 2$ and $\gamma/\omega_k = 0.2$.

features may sometimes arise, such as the appearance of band gaps at the mini-Brillouin-zone boundaries mentioned earlier. The properties for the occurrence and manipulation of such bands have led to the tremendous interest in *band-gap materials*, initially in the photonics case in optics (see [41] for a thorough review) but subsequently for other excitations as well. In this introductory chapter we present a simple formulation for a 1D photonic band-gap material (a photonic crystal) and discuss some extensions to 2D or 3D. This treatment will serve as a template for carrying out other calculations for periodic (and quasi-periodic) structures in the later chapters.

1.5.1 Bulk-slab model

Here we consider alternating layers of medium A and medium B to form the superlattice structure depicted in figure 1.8. The layer thicknesses are d_A and d_B, both assumed to be large compared with the atomic lattice parameters of the materials. The length associated with the artificial periodicity is $D = d_A + d_B$ and so the 1D wave vector Q (the so-called Bloch wave vector) in the z-direction associated with this translational symmetry has a first Brillouin zone corresponding to $-\pi/D < Q \leqslant \pi/D$. Further it will be assumed that the dielectric functions are ε_A and ε_B, taken for simplicity to be the bulk dielectric constants (independent of the excitation angular frequency ω). This is sometimes referred to as the *bulk-slab model* of a superlattice.

We now employ standard results from electromagnetism for the optical wave propagation in the layers, taken to be in the xz plane. There are basically two cases, depending on whether the waves have s-polarization (with electric-field vector **E** in the y-direction) or p-polarization (with **E** in the xz-plane). The wave-vector component Q enters into the calculations through Bloch's theorem in the form

$$\mathbf{E}(z + D) = \exp(\mathrm{i}QD)\mathbf{E}(z). \tag{1.46}$$

Also all field components will be taken to have dependences on x and t as $\exp(\mathrm{i}q_x x - \mathrm{i}\omega t)$, where for completeness we have included a wave vector q_x in the x-direction to allow for oblique incidence of the optical waves at each interface.

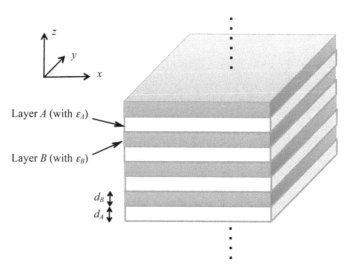

Figure 1.8. Geometry and notation for the calculation of the photonic (optical) band structure of a two-component alternating periodic superlattice. The wave propagation directions are taken to be in the xz plane.

The calculation for s-polarization will be shown first. The solution of the wave equation (1.19) for E_y in any layer will be a superposition of a forward- and a backward-travelling wave in the z-direction. Alternative forms can be written down, depending on whether the phases are taken relative to the lower (L) or upper (U) interface of each layer, giving

$$E_y = a_l^{\mathrm{L}} \exp\left[iq_{Az}(z - lD)\right] + b_l^{\mathrm{L}} \exp\left[-iq_{Az}(z - lD)\right]$$
$$= a_l^{\mathrm{U}} \exp\left[iq_{Az}(z - lD - d_A)\right] + b_l^{\mathrm{U}} \exp\left[-iq_{Az}(z - lD - d_A)\right] \qquad (1.47)$$

in an A layer (with $lD \leqslant z \leqslant lD + d_A$) where q_{Az} must satisfy

$$q_{Az}^2 + q_x^2 = \varepsilon_A \omega^2 / c^2. \qquad (1.48)$$

For the field in the adjacent layer B in the same cell l (with $lD + d_A \leqslant z \leqslant (l+1)D$) and with q_{Bz} defined similarly to equation (1.48), we have

$$E_y = d_l^{\mathrm{L}} \exp\left[iq_{Bz}(z - lD - d_A)\right] + e_l^{\mathrm{L}} \exp\left[-iq_{Bz}(z - lD - d_A)\right]$$
$$= d_l^{\mathrm{U}} \exp\left[iq_{Bz}(z - (l+1)D)\right] + e_l^{\mathrm{U}} \exp\left[-iq_{Bz}(z - (l+1)D\right]. \qquad (1.49)$$

Using a matrix notation the amplitudes in equation (1.47) are related by

$$\left|u_l^{\mathrm{U}}\right\rangle = F_A\left|u_l^{\mathrm{L}}\right\rangle, \qquad (1.50)$$

where we denote

$$\left|u_l^{\mathrm{L,U}}\right\rangle = \begin{pmatrix} a_l^{\mathrm{L,U}} \\ b_l^{\mathrm{L,U}} \end{pmatrix} \qquad (1.51)$$

and F_A is a 2×2 matrix given by

$$F_A = \begin{pmatrix} f_A & 0 \\ -0 & f_A^{-1} \end{pmatrix} \tag{1.52}$$

with phase term $f_A = \exp(iq_{Az}d_A)$. Likewise for the other layer we write

$$\left| w_l^U \right\rangle = F_B \left| w_l^L \right\rangle \tag{1.53}$$

with analogous notations and definitions.

The electromagnetic boundary conditions at the $z = lD + d_A$ and $z = (l + 1)D$ interfaces require that E_y and H_x are continuous (implying that E_y and $\partial E_y/\partial z$ are continuous). These can be used to give additional relationships between the column matrices of coefficients. The results can eventually be expressed in matrix form as

$$X_A \left| u_l^U \right\rangle = X_B \left| w_l^L \right\rangle \quad \text{and} \quad X_B \left| w_l^U \right\rangle = X_A \left| u_{l+1}^L \right\rangle, \tag{1.54}$$

where

$$X_i = \begin{pmatrix} 1 & 1 \\ q_{iz} & -q_{iz} \end{pmatrix}, \quad i = A, B. \tag{1.55}$$

The equations (1.50), (1.53) and (1.54) may now be combined to give

$$\left| u_{l+1}^L \right\rangle = T \left| u_l^L \right\rangle, \tag{1.56}$$

which relates amplitudes in cell l to the equivalent part in cell $l + 1$ and introduces the *transfer matrix* T given by

$$T = X_A^{-1} X_B F_B X_B^{-1} X_A F_A. \tag{1.57}$$

The transfer matrix has some useful properties, as we now explain. First, it is unimodular in the sense that $\det T = 1$, which can be proved from the form of matrix products in equation (1.57). Second, the statement of Bloch's theorem in equation (1.46) is equivalent to

$$\left| u_{l+1}^L \right\rangle = \exp(iQD) \left| u_l^L \right\rangle \tag{1.58}$$

and therefore we have the property that

$$\left[T - \exp(iQD)I \right] \left| u_l^L \right\rangle = 0, \tag{1.59}$$

where I is the unit 2×2 matrix. There is a similar equation obtained by relating $|u_l^L\rangle$ to $|u_l^L\rangle$, yielding

$$\left[T^{-1} - \exp(-iQD)I \right] \left| u_l^L \right\rangle = 0. \tag{1.60}$$

By adding equations (1.59) and (1.61) we find

$$\left[T + T^{-1} - 2\cos(QD)I \right] |u_l^L\rangle = 0, \tag{1.61}$$

which holds for any cell l. Therefore, from the property that $(T + T^{-1}) = (\mathrm{tr}\,T)I$, where 'tr' denotes the trace, we must have

$$\cos(QD) = \frac{1}{2}\,\mathrm{tr}\,T. \tag{1.62}$$

This is a general result, providing an implicit dispersion relation for the excitation frequencies in terms of the Bloch wave vector Q and the transfer matrix T.

Explicit evaluation shows that the diagonal elements of T in the present example are

$$T_{11} = f_A\left[f_B\left(q_{Az} + q_{Bz}\right)^2 - f_B^{-1}\left(q_{Az} - q_{Bz}\right)^2 \right] \Big/ 4q_{Az}q_{Bz},$$
$$T_{22} = f_A^{-1}\left[f_B^{-1}\left(q_{Az} + q_{Bz}\right)^2 - f_B\left(q_{Az} - q_{Bz}\right)^2 \right] \Big/ 4q_{Az}q_{Bz}. \tag{1.63}$$

When these results are substituted into equation (1.62) we find after some algebraic manipulation that the dispersion relation in this case of s-polarization is

$$\cos(QD) = \cos\left(q_{Az}d_A\right)\cos\left(q_{Bz}d_B\right) - g_s \sin\left(q_{Az}d_A\right)\sin\left(q_{Bz}d_B\right) \tag{1.64}$$

where

$$g_s = \frac{1}{2}\left(\frac{q_{Bz}}{q_{Az}} + \frac{q_{Az}}{q_{Bz}} \right). \tag{1.65}$$

The calculation for the case of p-polarization is very similar and it is found that equation (1.64) is still applicable provided g_s is replaced by g_p where

$$g_p = \frac{1}{2}\left(\frac{\varepsilon_B q_{Az}}{\varepsilon_A q_{Bz}} + \frac{\varepsilon_A q_{Bz}}{\varepsilon_B q_{Az}} \right). \tag{1.66}$$

The dispersion equation in the general form represented by (1.64) occurs commonly for excitations in periodic structures, as we will see in later chapters. It is sometimes called the *Rytov equation*, since its first derivation is attributed to Rytov in the context of acoustic waves [42]. Full discussion of the results for the above optical superlattice, or photonic crystal, can be found in [41, 43]. We present a numerical example of a dispersion relation in figure 1.9 in the special case of normal incidence where $q_x = 0$ and the distinction between s- and p-polarization vanishes. The plot is in dimensionless units for frequency ω versus Q. In the reduced-zone scheme used here the dispersion curves are 'folded back' and gaps open up at $QD = 0$ and π. In this example the band gaps (or stop bands) are relatively large because we have assumed a significant difference, or mismatch, between the dielectric constants

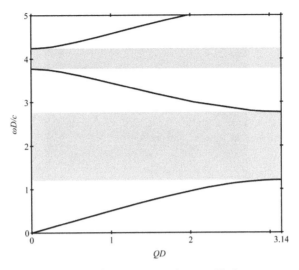

Figure 1.9. Dispersion curves of frequency (in terms of $\omega D/c$) versus Bloch wave vector (in terms of QD) for normal-incidence optical waves in a 1D photonic crystal consisting of a two-component alternating periodic superlattice. The assumed parameters are $d_A = d_B = 0.5D$, $\varepsilon_A = 3$ and $\varepsilon_B = 2$. The PBGs are shown as the shaded areas.

of adjacent layers. If the dielectric constants are taken to be close in value, the gaps shrink to a small value.

Finally we remark that the transfer-matrix method can straightforwardly be modified to apply other types of excitations, as well as to study the localized surface and/or interface modes (when they exist) in periodic and quasi-periodic structures that are either semi-infinite or finite in extent (in the z-direction). Results to illustrate this will be covered in some of the later chapters.

1.5.2 Photonic crystals in 2D and 3D

In the previous subsection it was demonstrated that stop bands would arise as a result of 1D periodicity in a superlattice. The interesting question arises as to whether it is possible to achieve 2D, or even 3D, arrays that would have stop bands or band gaps for optical signals at some frequencies and for both polarizations. The first investigations relating to this were made by Yablonovitch [23] in 1987, while related light localization studies were carried out around the same time by John [24]. A motivation for these early studies was the realization that a stop band with the properties mentioned would prohibit spontaneous emission of radiation in optical transitions and this could be utilized for semiconductor lasers and devices. However, subsequently a wide range of other quantum-optical device applications has emerged and an excellent reference is the book by Joannopoulos *et al* [41], which also covers the modelling techniques used to calculate the stop bands and the transmission characteristics.

The first realization of a full photonic band gap (PBG) material was due to Yablonovitch *et al* [44] who formed a structure that absorbed 15 GHz radiation for all spatial directions and polarizations. It was made by drilling air holes at various

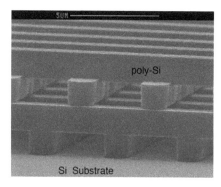

Figure 1.10. SEM of a 3D photonic crystal with a 'woodpile' structure. The rods are made of polycrystalline Si with widths 1.2 μm. The stacking sequence shown here repeats itself every four layers. Scale bar, 5 μm. After Lin *et al* [45].

angles into a dielectric block. Such an arrangement, whilst effective, would be impractical for scaling to the higher frequencies (shorter wavelengths) required for most applications. Versatile alternative structures have been based on the so-called 'woodpile' periodic arrangements of semiconductor rods, in which each rod may have a rectangular cross section with dimensions of order 1 μm and a length that is much larger. Such an arrangement using rods of polycrystalline silicon (dielectric constant ~13) on a Si substrate is illustrated in figure 1.10. In this structure the spatial alignments and separations between rods constitute an effective face-centred tetragonal lattice symmetry and the stop band corresponds to the 10–14.4 μm range of wavelengths. Another PBG structure that has been investigated (see [41]) is the 'inverse opal' arrangement. These are formed by utilizing the property of microscopic spheres (e.g., of silica, SiO_2) to self-assemble into an fcc lattice from a colloidal solution. The structure can then be inverted by filling the voids between spheres with a high-dielectric material and finally dissolving away the spheres.

As well as the intense activity mentioned above with the focus on 3D PBG materials, there have also been advances in developing 2D structures [41]. In particular, there have been studies of structures with artificial 2D periodicity, including cases where the dielectric functions are dispersive, giving rise to plasmons, phonons and polaritons, for example. We will discuss some of these 2D cases in the later chapters.

References

[1] Kittel C 2004 *Introduction to Solid State Physics* 8th edn (New York: Wiley)
[2] Ashcroft N W and Mermin N D 1976 *Solid State Physics* (New York: Saunders College)
[3] Grosso G and Parravicini G P 2013 *Solid State Physics* 2nd edn (Amsterdam: Academic)
[4] Cottam M G and Tilley D R 2005 *Introduction to Surface and Superlattice Excitations* 2nd edn (Bristol: IOP)
[5] Bloch F 1928 *Z. Phys.* **52** 555
[6] Novoselov K S, Geim A K, Morozov S V, Jiang D, Katsnelson M I, Grigorieva I V, Dubonos S V and Firsov A A 2005 *Nature* **438** 197

[7] Castro Neto A H, Guinea F, Peres N M R, Novosolov K S and Geim A K 2009 *Rev. Mod. Phys.* **81** 109

[8] Nénert G, Ritter C, Isobe M, Isnard O, Vasiliev A N and Ueda Y 2009 *Phys. Rev.* B **80** 024402

[9] Yariv A 1989 *Quantum Electronics* 3rd edn (New York: Wiley)

[10] Butcher P N and Cotter D 1990 *The Elements of Nonlinear Optics* (Cambridge: Cambridge University Press)

[11] Mills D L 1998 *Nonlinear Optics* 2nd edn (Berlin: Springer)

[12] Gurevich A G and Melkov G A 1996 *Magnetization Oscillations and Waves* (Boca Raton, FL: CRC)

[13] Stancil D D and Prabhakar A 2009 *Spin Waves: Theory and Application* (Heidelberg: Springer)

[14] Jackson J D 1999 *Classical Electrodynamics* 3rd edn (New York: Wiley)

[15] Griffiths D J 2012 *Introduction to Electrodynamics* 4th edn (Boston, MA: Addison-Wesley)

[16] Stegeman G I, Seaton C T, Hetherington W M, Boardman A D and Egan P 1986 *Electromagnetic Surface Excitations* ed R F Wallis and G I Stegeman (Berlin: Springer) p 261

[17] Wendler L 1986 *Phys. Stat. Solidi* B **135** 759

[18] Weisbuch B and Vinter B 1991 *Quantum Semiconductor Structures* (Boston, MA: Academic)

[19] Vollath D 2008 *Nanomaterials* (Weinheim: Wiley)

[20] Esaki L and Tsu R 1970 *IBM J. Dev.* **14** 61

[21] MacDonald A H 1987 *Interfaces, Quantum Wells, and Superlattices* ed C R Leavens and R Taylor (New York: Plenum) p 347

[22] Albuquerque E L and Cottam M G 2003 *Phys. Rep.* **376** 225

[23] Yablonovich E 1987 *Phys. Rev. Lett.* **58** 2059

[24] John S 1987 *Phys. Rev. Lett.* **58** 2486

[25] Mitin V V, Kochelap V A and Stroscio M A 2008 *Introduction to Nanoelectronics: Science, Nanotechnology, Engineering, and Applications* (Cambridge: Cambridge University Press)

[26] Aliofkhazraei M and Ali N 2012 *Two-Dimensional Nanostructures* (Boca Raton, FL: CRC)

[27] Cottam M G and Lockwood D J 1986 *Light Scattering in Magnetic Solids* (New York: Wiley)

[28] Cardona M and Güntherodt G (ed) 1989 *Light Scattering in Solids* vol 5 (Berlin: Springer)

[29] Dutcher J 1995 *Linear and Nonlinear Spin Waves in Magnetic Films and Superlattices* ed M G Cottam (Singapore: World Scientific) p 287

[30] Hillebrands B and Ounadjela K (ed) 2003 *Spin Dynamics in Confined Magnetic Structures* vol 2 (Berlin: Springer)

[31] Woodruff D P and Delchar T A 1986 *Modern Techniques of Surface Science* (Cambridge: Cambridge University Press)

[32] Feldman L C and Mayer J W 1986 *Fundamentals of Surface and Thin Film Analysis* (Amsterdam: North-Holland)

[33] Lehwald S, Szeftel J, Ibach H, Rahman T S and Mills D L 1983 *Phys. Rev. Lett.* **50** 518

[34] Hopster H 1994 *Ultrathin Magnetic Structures* vol 1 ed J A C Bland and B Heinrich (Berlin: Springer)

[35] Schreyer A, Schmitte T, Siebrecht R, Bödeker P, Zabel H, Lee S H, Erwin R W, Majkrzak C F, Kwo J and Hong M 2000 *J. Appl. Phys.* **87** 5443

[36] Brusdeylins G, Doak R B and Toennies J P 1981 *Phys. Rev. Lett.* **46** 437

[37] The Object Oriented MicroMagnetic Framework (OOMMF) project at ITL/NIST http://math.nist.gov/oommf

[38] Bransden B H and Joachain C J 2000 *Quantum Mechanics* 2nd edn (Englewood Cliffs, NJ: Prentice-Hall)

[39] Landau L D and Lifshitz E M 1980 *Statistical Physics* (Oxford: Pergamon)

[40] Cottam M G and Maradudin A A 1984 *Surface Excitations* ed V M Agranovich and R Loudon (Amsterdam: North-Holland) p 1

[41] Joannopoulos J D, Johnson S G, Winn J N and Meade R D 2008 *Photonic Crystals: Molding the Flow of Light* 2nd edn (Princeton, NJ: Princeton University Press)

[42] Rytov S M 1956 *Sov. Phys. Acoust.* **2** 68

[43] Yariv A and Yeh P 1984 *Optical Waves in Crystals* (New York: Wiley)

[44] Yablonovitch E, Gmitter T J and Leung K M 1991 *Phys. Rev. Lett.* **67** 2295

[45] Lin S Y, Fleming J G, Hetherington D L, Smith B K, Biswas R, Ho K M, Sigalas M M, Zubrzycki W, Kurtz S R and Bur J 1998 *Nature* **394** 251

IOP Publishing

Dynamical Properties in Nanostructured and
Low-Dimensional Materials

Michael G Cottam

Chapter 2

Phonons

In this and the next several chapters we will describe a range of different types of excitations in low-dimensional systems, starting here with phonons, or quantized vibrational excitations. We build on many of the concepts and methods introduced in chapter 1. The focus in the first instance will be on the *linear* properties of the excitations, but many of the results will be extended to the NL dynamics in chapter 7.

There are two main approaches followed in describing phonons in materials. One of these involves lattice dynamics for the discrete array of atoms in the material. This technique may be difficult to apply and is often idealized in the form of 1D models, but it has the advantage of producing results that are applicable for the entire Brillouin zone. The other approach is to deal with the regime of small wave vectors (or equivalently large wavelengths compared with the lattice constants of the material). Then a *continuous-medium approximation* can be adopted in which the underlying lattice structure is ignored and the macroscopic variables of elasticity theory are used. This leads to a simplified description of the long-wavelength acoustic phonons, but it does not account for the higher-frequency optic phonons. In this chapter we will follow both approaches and both will be employed in the interpretation of experimental data.

General accounts of phonons are to be found in textbooks on condensed-matter physics (see, e.g. [1, 2]) and in more specialized books such as [3, 4]. Some references that emphasize applications to surfaces are [5–7]. Here we will emphasize applications to phonons in a variety of low-dimensional structures, including periodic and quasi-periodic arrays of nanoelements and recent developments for phononic crystals. It is worth noting that most of the results required in this chapter for phonons in the linear regime can be obtained using classical (rather than quantum-mechanical) methods. Some general references that give more emphasis to the latter approach are [3, 4].

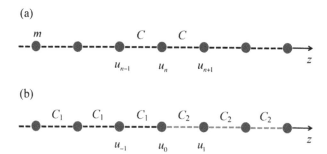

Figure 2.1. Models in 1D used for the lattice dynamics of monatomic chains, denoting the longitudinal displacement of atom n by u_n: (a) a simple chain of equally spaced masses m with the same spring constant C providing nearest-neighbour coupling and (b) an interface between two chains where the masses are all the same but the spring constants are different on the right and left of the interface mass.

2.1 Lattice dynamics for single surfaces and films

2.1.1 Monatomic chains

We begin with examples for 1D chains of atoms and then generalize to more complex structures in one and higher dimensions. The simplest case is a 1D infinite lattice of identical atoms located at $z = na$ (with integers $-\infty < n < \infty$) along the z-axis as in figure 2.1(a). Assuming that each atom has mass m and is coupled to its two nearest neighbours by an elastic spring with force constant denoted by C, Newton's equation of motion for the longitudinal displacement u_n of the nth atom is

$$m\frac{d^2 u_n}{dt^2} = C[(u_{n+1} - u_n) - (u_n - u_{n-1})].\qquad(2.1)$$

This equation can be solved by seeking normal-mode solutions with angular frequency ω and a spatial dependence in accordance with Bloch's theorem as in equation (1.29), namely $u_n \propto \exp[i(qna - \omega t)]$, yielding

$$\omega^2 = \frac{2C}{m}[1 - \cos(qa)] = \frac{4C}{m}\sin^2(qa/2).\qquad(2.2)$$

A plot of this dispersion relation is shown in figure 2.2 for the first Brillouin zone, which extends from $q = -\pi/a$ to $q = \pi/a$. Some limiting cases are $\omega = v_L|q|$ when $|qa| \ll 1$, where we denote $v_L = \sqrt{Ca^2/m}$ and then $\omega = 2\sqrt{C/m}$ when $|qa| = \pi$. The first case is for the regime of long wavelengths (compared to a) where we have essentially a sound wave propagating longitudinally with speed v_L and the discrete-lattice structure has become unimportant. The second case represents the maximum phonon frequency, which corresponds to a standing wave at the Brillouin-zone boundary.

The above simple calculation can be generalized to include effects of surfaces and interfaces. First we consider what happens if the monatomic chain in figure 2.1(a) is semi-infinite, i.e., it has a termination point at one end (on the left) such that $n = 1, 2, ..., \infty$. Bloch's theorem will not apply, but we may still seek normal-mode solutions with time dependence such as $\exp(-i\omega t)$ whereupon we have

$$-m\omega^2 u_1 = C(u_2 - u_1), \quad (n = 1),$$
$$-m\omega^2 u_n = C(u_{n+1} - 2u_n + u_{n-1}), \quad (n \geqslant 2).\qquad(2.3)$$

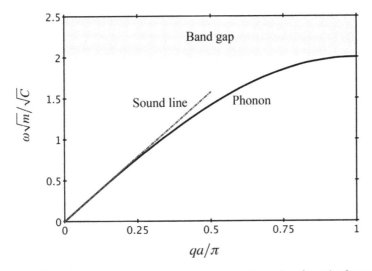

Figure 2.2. Phonon dispersion relation for a monatomic chain showing a plot of angular frequency ω versus 1D wave vector for q ranging from 0 to the first Brillouin-zone boundary at π/a, according to equation (2.2). The tangential line at the origin represents the dispersion for a sound wave.

The first line above represents a modified equation (a boundary condition) for the 'surface' atom at the end of the chain. It can be shown, e.g., by direct substitution, that the solutions for the displacements are of the form

$$u_n = f \exp(iqna) + g \exp(-iqna), \tag{2.4}$$

where f and g are constants independent of n. This is just a superposition of waves travelling to the right and left with amplitudes f and g, respectively, and application of the boundary equation for $n = 1$ yields an expression for the reflection coefficient corresponding to f/g. In this case it is found that no solutions exist for the lattice dynamics corresponding to localized modes with decay-like characteristics from the free end (see additional discussion in the appendix).

Next we briefly examine the situation depicted in figure 2.1(b) which models the interface in 1D between two media. We assume a fully infinite chain where the masses are all the same, but the spring constant changes abruptly from C_2 to C_1 at one of the masses labelled as $n = 0$. The equations of motion for the masses with $n < 0$ and $n > 0$ have the same form as equation (2.1) when the appropriate value of the spring constant has been inserted, but the equation for the interface ($n = 0$) mass has the modified form

$$m\frac{d^2u_0}{dt^2} = C_2(u_1 - u_0) - C_1(u_0 - u_{-1}). \tag{2.5}$$

The solutions to the left and right of the interface are found to take the same form as discussed above, as in equation (2.4), and again there are no localized or spatially decaying modes. It is useful for later applications to consider the form of the interface condition in a continuum approximation. The finite differences in equation (2.5) may be replaced by spatial derivatives giving

$$\rho a \frac{\partial^2 u}{\partial t^2} = K_2 \frac{\partial u_+}{\partial z} - K_1 \frac{\partial u_-}{\partial z}, \tag{2.6}$$

where $\rho = m/a$ is a density and $K = Ca$ is an elastic modulus, while u_- and u_+ denote the continuum displacements to the left and right of the interface. This results simplifies further if we observe that the left-hand side is of second order in the small quantity qa and so can be neglected compared with the right-hand side of order qa, giving

$$K_2 \frac{\partial u_+}{\partial z} = K_1 \frac{\partial u_-}{\partial z} \tag{2.7}$$

for the interface condition.

Our final remarks on monatomic chains will be applied to the situation where the chain shown in figure 2.2(a) is terminated at both ends, so that $n = 1, 2, ..., N$ for a chain of length $L = (N - 1)a$. Most of the mathematics proceeds as before using equation (2.4) for the displacement u_n. However, we now obtain two expressions for the ratio f/g, one when $n = 1$ and another when $n = N$. Equating them provides us with a consistency condition to be satisfied by qa, the length L and the spring constant C. The actual form of this condition is unimportant here and the main point is that for a given L and C it will be satisfied only for certain discrete values of qa (in fact, there are N values as expected for the N normal modes of vibration for the system). This contrasts with the situation for infinite or semi-infinite chains where qa was unrestricted. This 'quantization' of the wave vector frequently occurs in finite systems, as we will see throughout later examples.

2.1.2 Diatomic chains

We now extend the calculations in the previous subsection to the more interesting case of a diatomic chain modelled as in figure 2.3(a) with alternating masses denoted by m_1 and m_2 and with the nearest neighbours coupled by identical springs with elastic constant C. If the spacing between the atoms is a, it follows that the periodicity length for an infinitely extended chain is $2a$ so the 1D Brillouin zone in the z-direction has a wave-vector component q extending from $-\pi/2a$ to $\pi/2a$. We take the masses to be located at $z = (2n + 1)a$ and $z = 2na$ for the m_1 and m_2 sites, respectively, where integers $-\infty < n < \infty$. By analogy with the monatomic-chain case, and seeking normal modes with a time dependence such as $\exp(-i\omega t)$, we now have two sets of coupled equations:

$$\begin{aligned}
-m_1\omega^2 u_{2n+1} &= C(u_{2n+2} - 2u_{2n+1} + u_{2n}), \\
-m_2\omega^2 u_{2n} &= C(u_{2n+1} - 2u_{2n} + u_{2n-1}).
\end{aligned} \tag{2.8}$$

The solution for the infinitely extended system is easily found using Bloch's theorem, which implies here that $u_{2n+1} = U_1 \exp[iq(2n + 1)a]$ and $u_{2n} = U_2 \exp[iq2na]$, where amplitude terms U_1 and U_2 are independent of the spatial variables. Substitution into

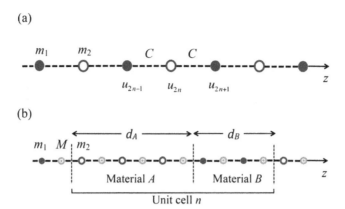

Figure 2.3. Model in 1D used for the lattice dynamics of diatomic chains, where the notations are similar to figure 2.1: (a) a simple diatomic chain with alternating masses m_1 and m_2 and with the spring constant C taken to be the same between all nearest neighbours and (b) model for a diatomic superlattice with masses denoted by m_1, m_2 and M (see the text).

equation (2.8) leads to two expressions for the ratio U_1/U_2, which may then be equated to obtain the phonon dispersion relation in the form of a quadratic equation in ω^2. This may be solved to give the usual textbook result [1, 2] that

$$\omega^2 = C\left(\frac{1}{m_1} + \frac{1}{m_2}\right) \pm C\left[\left(\frac{1}{m_1} + \frac{1}{m_2}\right)^2 - \frac{4\sin^2(qa)}{m_1 m_2}\right]^{1/2}. \tag{2.9}$$

This consists of two distinct branches, as might be expected since there are two masses in the unit cell. Taking the upper sign, we have an *optic phonon branch* in a frequency range $(2C/m_2)^{1/2} \leqslant \omega \leqslant [2C(m_1 + m_2)/m_1 m_2]^{1/2}$ assuming $m_1 > m_2$, whereas taking the lower sign, we have an *acoustic-phonon branch* in a frequency range $0 \leqslant \omega \leqslant (2C/m_1)^{1/2}$. There is an intervening region with frequencies from $(2C/m_1)^{1/2}$ up to $(2C/m_2)^{1/2}$ where there are no phonon excitations and this is referred to as a *stop band* or *band gap*. The phonon dispersion relations are illustrated in figure 2.4.

The above-mentioned stop band and the higher-frequency band where $\omega > [2C(m_1 + m_2)/m_1 m_2]^{1/2}$ are where localized modes may be found to occur in the related diatomic chains obtained if there are surfaces and interfaces to remove the translational symmetry in the z-direction. To illustrate this point we consider the case where the diatomic chain terminates at the left-hand end with a mass m_2 at $n = 0$. The coupled equations for all $n > 0$ are formally the same as in (2.8), whereas the 'surface' $n = 0$ mass satisfies the boundary equation

$$-m_2\omega^2 u_0 = C(u_1 - u_0). \tag{2.10}$$

There are now two possible ways that we could proceed with the calculation. One way is to assume the same form of solutions as we employed following equation (2.8), except that solutions with q complex (as well as real) become allowable since Bloch's theorem is no longer applicable. For localized solutions that decay with distance away

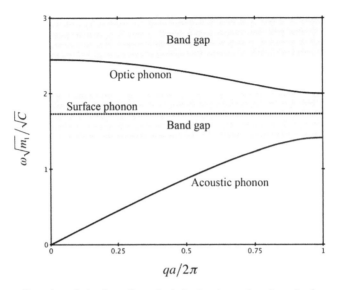

Figure 2.4. Phonon dispersion relation for a diatomic chain showing a plot of angular frequency ω versus 1D wave vector for q ranging from 0 to the first Brillouin-zone boundary at $\pi/2a$, according to equation (2.9). The acoustic and optic phonon branches are shown, as well as the dispersionless surface phonon branch for a semi-infinite chain, which occurs within the lower band gap (see equation (2.11) and the text).

from the surface (as n becomes large) we require $\text{Im}(q) > 0$. It is then straightforward to show that there is a surface phonon solution with angular frequency given by

$$\omega_S = \left[C \left(\frac{1}{m_1} + \frac{1}{m_2} \right) \right]^{1/2}, \qquad (2.11)$$

that exists only when $m_2 < m_1$, i.e., the chain terminates with the lighter mass. It occurs within the stop band between the acoustic and optic phonons (see figure 2.4). The other way to arrive at the same result more rigorously, i.e., without making assumptions about the form of the solution, is to follow the TDM method as outlined in section A.1 of the appendix.

Various extensions may be made of the calculations for diatomic chains. One possibility is to consider a finite-length chain with N atoms. If N is even, then one termination atom will necessarily be with the heavier mass and the other one will be with the lighter mass. It is found in this case that a surface mode occurs at just one end (with the lighter mass), as expected, and there are $N - 1$ other normal modes in the form of quantized bulk modes. If N is odd we may have either two surface modes or no surface modes, depending on the termination conditions. A careful analysis of this problem was given by Wallis [8]. Another extension would be to consider an interface between two diatomic chains, but we will defer this topic until section 2.4 on superlattices where we examine the behaviour in the context of multiple interfaces.

Throughout this section, in considering both monatomic and diatomic 1D chains, we are looking at systems where there is no wave vector $\mathbf{q}_\parallel$ in a surface plane, as would be the case more generally in solids. In principle, the lattice-dynamics approach can be extended to 2D and 3D (see, e.g., [9, 10]); further references and discussion are to be found in the review article by Esser and Richter [11]. The calculations, however, rapidly become laborious unless numerical simulation packages are employed. An example of a more recent investigation, which was applied to MoS_2 in both its 2D single-layer and 3D multilayered forms, can be found in [12].

2.2 Elastic waves for single surfaces and films

An introductory example involving elastic waves was given in section 1.4, taking the case of an infinitely extended isotropic medium. We now develop this approach more formally and with the effects of surfaces and interfaces included. Some simple cases of elasticity are covered in general books on solid-state physics (see, e.g., [1, 2]), while a more detailed account can be found in [13].

The starting point is to introduce the *strain* $\bar{u}_{ij}$ defined in terms of the displacement $\mathbf{u}(\mathbf{r}, t)$ at position $\mathbf{r}$ and time t by

$$\bar{u}_{ij} = \frac{1}{2}\left(\frac{\partial u_i}{\partial r_j} + \frac{\partial u_j}{\partial r_i}\right), \tag{2.12}$$

where u_i and r_i, with the i and j labels denoting 1, 2, 3 (or equivalently x, y, z), are the components of vectors $\mathbf{u}$ and $\mathbf{r}$. Next the *stress* σ_{ij} is defined by the relationship

$$\delta F_i = \sum_k \sigma_{ik}\hat{n}_k \delta s, \tag{2.13}$$

where $\delta \mathbf{F}$ is the force on an elementary surface area δs normal to the unit vector $\hat{\mathbf{n}}$. In general, both the stress and strain are second-rank tensors (or matrices) and they are symmetric, so $\sigma_{ij} = \sigma_{ji}$, etc. They are linearly related through *Hooke's law*, which states that

$$\sigma_{ij} = \sum_{kl} \lambda_{ijkl}\bar{u}_{kl}. \tag{2.14}$$

Here λ_{ijkl} is the *elasticity tensor*. It is of fourth rank and symmetric under interchange of labels i and j and of k and l, as well as satisfying $\lambda_{ijkl} = \lambda_{klij}$. From these symmetries it may be inferred that the elasticity tensor may have up to 21 independent components. For isotropic media, however, it can be shown that there are actually only two independent components and these are conventionally expressed in terms of Young's modulus E and Poisson's ratio σ. The stress–strain relationships then take the form

$$\sigma_{xx} = \frac{E}{(1 + \sigma)(1 - 2\sigma)}\left[(1 - \sigma)\bar{u}_{xx} + \sigma(\bar{u}_{yy} + \bar{u}_{zz})\right],$$
$$\sigma_{xy} = \frac{E}{(1 + \sigma)}\bar{u}_{xy}, \tag{2.15}$$

with analogous expressions for σ_{yy}, σ_{yz}, etc. We note that typically $0 \leqslant \sigma \leqslant 1/2$.

Using the above definitions, the equation of motion of an element of mass can be expressed in the form (see [13])

$$\rho\frac{\partial^2 \mathbf{u}}{\partial t^2} = \frac{E}{2(1+\sigma)}\nabla^2\mathbf{u} + \frac{E}{2(1+\sigma)(1-2\sigma)}\nabla(\nabla\cdot\mathbf{u}), \qquad (2.16)$$

where ρ is the density. There is a simplification in the cases of waves that propagate either longitudinally (with wave vector $\mathbf{q}$ parallel to the displacement denoted as $\mathbf{u}_L$) or transversely (with $\mathbf{q}$ perpendicular to the displacement $\mathbf{u}_T$). Then there is a separation into two wave equations, one with the form

$$\rho\frac{\partial^2 \mathbf{u}_L}{\partial t^2} = v_L^2\nabla^2\mathbf{u}_L \qquad (2.17)$$

with the longitudinal velocity v_L given by

$$v_L^2 = \frac{E(1-\sigma)}{\rho(1+\sigma)(1-2\sigma)}. \qquad (2.18)$$

Also there is a similar wave equation satisfied by the transverse displacement $\mathbf{u}_T$, but now the velocity v_T is given by

$$v_T^2 = \frac{E}{2\rho(1+\sigma)}. \qquad (2.19)$$

It follows from the above expressions and from the inequality mentioned for σ that $v_L \geqslant \sqrt{2}\,v_T$. As in subsection 1.4.1 the dispersion relations for the two types of bulk acoustic waves (without damping) are $\omega = v_L q$ and $\omega = v_T q$.

The other basic properties that we will require relate to the conditions satisfied by the stress and strain at any boundary. As the simplest example, we start by considering the reflection of bulk acoustic waves at a planar surface, taken as the plane $z = 0$ with the isotropic elastic medium filling the half-space $z < 0$. If the surface plane is assumed to be stress-free, then $\sigma_{iz} = 0$ (for $i = x, y, z$), which implies using equation (2.15) that

$$\bar{u}_{xz} = 0, \qquad\qquad \bar{u}_{yz} = 0,$$
$$\sigma(\bar{u}_{xx} + \bar{u}_{yy}) + (1-\sigma)\bar{u}_{zz} = 0. \qquad (2.20)$$

To illustrate these boundary conditions we consider, as in [7], the problem of solving for the reflected signal when a *transverse* bulk acoustic wave is incident on the planar surface from the region $z < 0$. The case of a *longitudinal* bulk acoustic wave can be treated in an analogous fashion, but will not be presented. If the incident plane (as defined by $\mathbf{q}$ and the normal to the surface) is taken to be the xz-plane, then there are two distinct cases arising for a transverse wave depending on the polarization of the displacement vector $\mathbf{u}$. In one case (known as s-polarization) $\mathbf{u}$ is in the y-direction perpendicular to the incident plane and in the other case (known

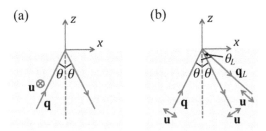

Figure 2.5. Reflection of a transverse bulk acoustic wave at a planar free surface ($z = 0$). The isotropic elastic medium fills the lower half-plane ($z < 0$), while the region $z > 0$ is vacuum. The two cases are for (a) s-polarization and (b) p-polarization.

as p-polarization) **u** is in the incident plane, as illustrated in figure 2.5. The incident and reflected waves in the s-polarized case are described by

$$\mathbf{u} = (0,\, a,\, 0)\exp\!\big[i(q_x x + q_z z)\big]\exp(-i\omega t) \quad \text{(incident)},$$
$$\mathbf{u} = (0,\, b,\, 0)\exp\!\big[i(q_x x - q_z z)\big]\exp(-i\omega t) \quad \text{(reflected)},$$
$$(2.21)$$

where a and b are the incident and reflected amplitudes, respectively. It follows from the relationships $\tan\theta = q_x/q_z$ and $\omega^2 = v_\mathrm{T}^2(q_x^2 + q_z^2)$ that the q_z values are of the same magnitude and the angles of incidence and reflection are the same ($=\theta$) as indicated in figure 2.5(a). Also, by using the boundary conditions in equation (2.20), we conclude that $b = a$.

It soon becomes evident that the case of p-polarization is more complicated. Suppose we were to start in the same manner by writing down incident and reflected waves, which now would have the form

$$\mathbf{u} = (a\cos\theta,\, 0,\, -a\sin\theta)\exp\!\big[i(q_x x + q_z z)\big]\exp(-i\omega t) \quad \text{(incident)},$$
$$\mathbf{u} = (b\cos\theta,\, 0,\, -b\sin\theta)\exp\!\big[i(q_x x - q_z z)\big]\exp(-i\omega t) \quad \text{(reflected)},$$
$$(2.22)$$

with a and b again representing amplitudes. It becomes obvious, however, that this is an insufficient description of the reflection process, because there are two boundary conditions to be satisfied from equation (2.20) but so far only one reflected amplitude. The problem is resolved by noting that another reflected wave, having longitudinal character and a different propagation angle, is additionally allowed, as shown in figure 2.5(b). This longitudinal wave (with amplitude denoted by c) can be expressed as

$$\mathbf{u} = (c\sin\theta_\mathrm{L},\, 0,\, -c\cos\theta_\mathrm{L})\exp\!\big[i(q_x x - q_z z)\big]\exp(-i\omega t). \quad (2.23)$$

It is straightforward to deduce that

$$\frac{\sin\theta_\mathrm{L}}{\sin\theta} = \frac{v_\mathrm{L}}{v_\mathrm{T}} > 1, \quad (2.24)$$

so $\theta_\mathrm{L} > \theta$. We note that if the angle of incidence θ exceeds a critical value θ_c satisfying $\sin\theta_\mathrm{c} = v_\mathrm{T}/v_\mathrm{L}$, then equation (2.24) would imply a value for $\sin\theta_\mathrm{L}$ greater than unity. The propagation equations can then be shown to require $q_x^2 > \omega^2/v_L^2$, which can only be possible if q_L^2 is negative. If we put $q_L = i\kappa_L$, where κ_L is real and positive, then equation (2.23) for the longitudinal wave is modified to take the form

$$\mathbf{u} = (c\sin\theta_L, 0, -c\cos\theta_L)\exp\left[i(q_x x)\exp(\kappa_L z)\right]\exp(-i\omega t). \tag{2.25}$$

We now recognize this as a wave travelling in the x-direction parallel to the surface and decaying with distance into the elastic medium. In these circumstances the reflected signal is sometimes referred to as a pseudo-surface wave. Relationships between amplitudes a, b and c can straightforwardly be deduced (e.g., by consideration of the energy flux).

Before proceeding to calculations of surface and interface modes in a more general context, we note in passing that the linear-response calculation of a displacement–displacement Green function given in subsection 1.4.2 for an infinite elastic medium can be extended to apply to the longitudinal and transverse acoustic (TA) modes in the semi-infinite medium case depicted in figure 2.5. A detailed discussion can be found in [10].

2.2.1 Planar surfaces and films

For the first example of an elastic surface wave we again consider a planar free surface as in figure 2.5 to derive the dispersion relation for a *Rayleigh wave*. It is a surface wave that can be formed from a mixture of waves with p-transverse and longitudinal components. The total displacement of a wave localized in the lower half-plane $z < 0$ can be expressed as the sum $\mathbf{u}_L + \mathbf{u}_T$, where the longitudinal and transverse parts must each satisfy the wave equation and have z-dependences such as $\exp(\kappa_{L,T})$ with $\kappa_L > 0$ and $\kappa_T > 0$. The in-plane component q_x of the wave vector must be the same for each term in order for the boundary conditions to be satisfied for all x. By using the wave equation we deduce that

$$\kappa_{L,T} = \left(q_x^2 - \omega^2/v_{L,T}^2\right)^{1/2}. \tag{2.26}$$

To proceed with the calculation for the properties of the surface wave, we first note that the solutions for $\mathbf{u}_L$ and $\mathbf{u}_T$ must have the form

$$\mathbf{u}_L = \left(aq_x, 0, -ia\kappa_L\right)\exp\left(iq_x x\right)\exp(\kappa_L z)\exp(-i\omega t),$$
$$\mathbf{u}_T = \left(b\kappa_T, 0, -ibq_x\right)\exp\left(iq_x x\right)\exp(\kappa_T z)\exp(-i\omega t), \tag{2.27}$$

in order that the conditions $\nabla \times \mathbf{u}_L = 0$ and $\nabla \cdot \mathbf{u}_T = 0$ can be satisfied. A relationship between the constants a and b can be found through application of the first boundary condition in equation (2.20) as

$$2\kappa_L q_x a + \left(q_x^2 + \kappa_T^2\right)b = 0. \tag{2.28}$$

The second boundary condition is automatically satisfied here, while the third eventually gives

$$\left(q_x^2 + \kappa_T^2\right)a + 2\kappa_T q_x b = 0 \tag{2.29}$$

after using the definitions of κ_L and κ_T, as well as equations (2.18) and (2.19). After eliminating a and b between equations (2.28) and (2.29) we arrive at an implicit dispersion relation in the form

$$\left(1 - \frac{\omega^2}{q_x^2 v_L^2}\right)\left(1 - \frac{\omega^2}{q_x^2 v_T^2}\right) = \left(1 - \frac{\omega^2}{2q_x^2 v_T^2}\right)^4. \tag{2.30}$$

Although this result looks complicated, it is just a polynomial in ω/q_x, so its solution can be written simply as

$$\omega = \xi v_T q_x, \tag{2.31}$$

where ξ is a dimensionless constant with a value that depends on v_T/v_L (or equivalently on Poisson's ratio σ). In fact, numerical solution of equation (2.30) shows that for the physical range of the parameters ξ is less than unity (and typically is close to 0.9). Details can be found in [7]. The conclusion from the above is that, for a given value of q_x, the frequency of the Rayleigh surface wave falls below that of the longitudinal and transverse bulk elastic waves.

Next, it is interesting to consider the elastic waves that can exist in a film of finite thickness L. This can be formed if a second free surface is introduced at $z = -L$ in figure 2.5. The elastic medium now occupies the region of space where $-L < z < 0$ and there are additional boundary conditions (ABCs) to implement at the second surface. There are two notable features regarding the elastic waves in this case. One is that a Rayleigh-type mode can be shown to exist at each surface. These two modes are essentially degenerate in frequency for large enough thicknesses (i.e., when $\kappa_L L \gg 1$ and $\kappa_T L \gg 1$), but for smaller thicknesses the modes form symmetric and antisymmetric combinations (in terms of the mode displacements across the film) and the frequencies are shifted and split. The general results are quite complicated but an approximate expression giving the Rayleigh-wave frequencies in a film of moderate thickness is quoted in [7].

The other modification for a film is the occurrence of modes where the z-component of a wave vector is real but quantized with discrete values (depending on the boundary conditions), in a similar manner to the quantization for a finite monatomic chain in subsection 2.1.1 except that we now have the wave polarization to take into account. The simplest case is for s-transverse polarization, where the solution for the y-component of the displacement has the form of a standing wave with respect to its z-dependence:

$$u_y = \left[a_1 \exp(iq_z z) + a_2 \exp(-iq_z z)\right] \exp(iq_x x) \exp(-i\omega t), \tag{2.32}$$

where q_z is real and satisfies $q_x^2 + q_z^2 = \omega^2/v_T^2$. Application of the boundary conditions, as in equation (2.20), allows the coefficients a_1 and a_2 to be eliminated to yield a dispersion relation. The situation with the coupled p-transverse and longitudinal modes is more complex and can give rise to what are often called *slab modes* in acoustics (or analogously *guided waves* in optics). Briefly, we could form solutions similar to the combinations in equation (2.27), except that some or all of the $\exp(\kappa_{L,T}z)$ terms for the z-dependence would be replaced by factors such as $\exp(iq_{L_z}z)$ and $\exp(iq_{T_z}z)$ with real wave-vector components. In this way it is possible to have mixtures of waves in which, for example, the transverse part is like a standing wave and the longitudinal part is like a localized wave at the surfaces. Collectively, all the elastic waves of a film are often referred to as *Lamb waves*.

So far we have considered only stress-free surfaces, but further possibilities for localized modes will arise if we extend some of our results to interfaces between two different elastic media. We mention some examples below. One case is the direct extension of the Rayleigh wave to a planar interface, where it is found that a localized mode called a *Stoneley wave* can exist provided the elastic constants of the two media satisfy certain criteria. Its dispersion relation can be written in the same form as equation (2.31), but the expression for ξ is more complicated (see, e.g., [14]). The Stoneley wave involves coupled p-transverse and longitudinal components, while the simplest example of an interface-related mode in s-polarization is the *Love wave*. This occurs for an elastic film of thickness L on a semi-infinite substrate of another material, provided the TA velocity v_T' of the substrate exceeds the value v_T in the film. This condition allows solutions to be found for displacements that are oscillatory functions of z inside the film and decaying functions of z with distance into the substrate [15].

The analysis of surface elastic waves in situations where there are *non-parallel* planar surfaces (as in edges or wedges) is surprisingly complicated, mainly because of the nature of the stress–strain relationships in even the most straightforward cases, e.g., see equation (2.15). Notably the elastic modes localized at a 90° edge were studied by Maradudin *et al* [16]. The modes are characterized by a 1D wave vector for propagation parallel to the edge and a decaying amplitude with respect to the two directions away from the planar surfaces. The decay properties were expressed in terms of Laguerre polynomials and it was found that the lowest-frequency edge mode has a slightly lower acoustic speed than that for the Rayleigh surface mode in equation (2.31). Subsequently generalizations were made to edges (or wedges) with other angles between the planar surfaces (see [5] for a review).

2.2.2 Cylinders and spheres

Here we consider excitations in systems with smooth curved surfaces, taking for simplicity the cases of spheres and long cylinders with a circular cross section. These provide us with examples of 0D nanodots and 1D nanowires, respectively (see figure 1.4).

The study of acoustic waves in a solid sphere of an isotropic elastic medium dates back to 1882 with calculations by Lamb [17] for the so-called spheroidal and torsional modes. A clear derivation of the main results is given in [18], where the starting point is the elastic wave equation (2.16) in its general vector form. On substituting for the ∇^2 operator in terms of spherical polar coordinates (r, θ, ϕ), as conventionally defined (see, e.g., [19]), with r as the radial distance from the centre of the sphere, θ as the polar angle and ϕ as the azimuthal angle, we obtain coupled partial differential equations satisfied by the components of the displacement vector.

The method of separation of variables can then be employed to find solutions that are linear combinations of products such as

$$j_\ell(\omega r/v_{L,T})Y_\ell^m(\theta, \phi)\exp(-i\omega t). \qquad (2.33)$$

Here j_ℓ is a spherical Bessel function of the first kind, Y_ℓ^m is a spherical harmonic, ω is an angular frequency, and ℓ and m are the usual quantum numbers associated with the angular momentum [19]:

$$\ell = 0, 1, 2, \ldots, \qquad m = -\ell, -\ell + 1, \ldots, \ell - 1, \ell. \qquad (2.34)$$

The mode frequencies are deduced by constructing solutions of the form (2.33) that satisfy the stress-free boundary conditions at $r = R$ for all values θ and ϕ, where R is the radius of the sphere. Two types of solutions arise as follows [18]: either

$$\frac{\eta j_{\ell+1}(\eta)}{j_\ell(\eta)} - (\ell - 1) = 0 \quad (\ell \geqslant 1) \qquad (2.35)$$

or

$$[\eta^2 - 2\ell(\ell - 1)(\ell + 2)]\frac{\eta j_{\ell+1}(\eta)}{j_\ell(\eta)} + \frac{1}{2}\eta^4 + (\ell - 1)(2\ell + 1)\eta^2$$

$$+ 2\left[\eta^2 + (\ell - 1)(\ell + 2)\left(\frac{\eta j_{\ell+1}(\eta)}{j_\ell(\eta)} - (\ell + 1)\right)\right]\frac{\xi j_{\ell+1}(\xi)}{j_\ell(\xi)} = 0 \quad (\ell \geqslant 0), \qquad (2.36)$$

where $\eta = \omega R/v_T$ and $\xi = \omega R/v_L$ are parameters that depend on the frequency. Equation (2.35) describes toroidal oscillations without dilation, where the displacement vectors are always parallel to the surface. Equation (2.36) describes spheroidal oscillations with dilation, i.e., the oscillations involve volume expansions and contractions of the sphere. As expected, the expressions for the mode frequencies have no wave-vector dependence, since the spheres are 0D structures. Also it can be noted that the equations for the frequencies are independent of the quantum number m, so each mode has a degeneracy of $2\ell + 1$. For each value of ℓ there are multiple solutions that may be labelled as $\omega_{n,\ell}$, where integer n ($= 1, 2, \ldots$) labels the modes for a given ℓ starting with the lowest frequency. There will be further discussion of these results in the next section in the context of Brillouin light scattering (BLS) experiments.

Turning now to the case of acoustic waves in a solid cylinder of an isotropic elastic medium, we mention in particular the calculations by Hudson [20] for modes propagating along the symmetry axis defined by the length of the cylinder. Later generalizations [21, 22] were made for Rayleigh-type modes propagating circumferentially. Consistent with this geometry it is appropriate to use cylindrical polar coordinates (r, θ, z), where z refers to the cylindrical axis while r and θ are the usual circular polar variables in the xy-plane. The solutions of the elastic wave equation become expressible in terms of solutions that are linear combinations of products such as

$$f(r)\exp(\pm im\theta)\exp(iq_z z)\exp(-i\omega t). \qquad (2.37)$$

Here the radial function $f(r)$ may involve various types of Bessel functions, m is an integer ($= 0, 1, 2, \ldots$) which ensures single-valuedness and q_z is a propagation wave vector along the length of the cylinder. The solutions for the allowed values of the angular frequency ω are found after applying the stress-free boundary conditions at all points on the curved surface at $r = R$. In general, there are multiple solutions for any value of m, so the modes can be labelled with an addition integer n as $\omega_{n,m}$. The results are described in [20–22]. An interesting property is that Rayleigh-type surface waves exist only if $q_z R$ exceeds a cut-off value, otherwise they are no longer localized and decay into bulk elastic waves.

2.3 Experimental studies

As described briefly in the introductory survey of experimental techniques in section 1.3, the inelastic scattering of light, by means of either BLS or RS, can provide a surface-sensitive technique with good resolution for detecting the surface and interface-related modes, particularly if relatively opaque materials are employed. A comprehensive review of RS studies for surface phonons has been given by Esser and Richter [11], so we will include just a few key examples here. Our focus in this section will be on summarizing the more extensive results available from BLS and particle scattering, where the latter allows for studies across the entire Brillouin zone of excitation wave vectors (see also [7]).

The surface RS technique has been used extensively to study materials (often III–V semiconductors such as GaAs or InP) with an ultrathin film of another material (such as Sb) deposited on their planar surface [11]. Typically the film might have a thickness of only one or a few MLs, or sometimes might have sub-ML coverage. An example is given in figure 2.6, which shows the RS spectra obtained for a InP sample when different thicknesses of Sb ranging from fractional coverage to one ML are deposited on the surface [23]. The observed evolution of the surface

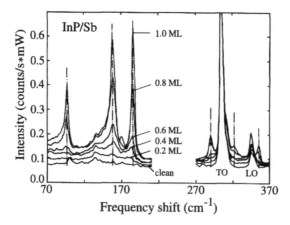

Figure 2.6. Spectra of intensity versus frequency shift for RS from phonons at the planar surface of InP/Sb samples. Results are shown for different thicknesses of Sb, expressed in MLs, on an InP (110) surface as the substrate. The transverse optic and longitudinal optic phonon peaks of InP are indicated, as well as the surface phonons of the Sb layer (vertical dash-dot lines). After Hünermann *et al* [23].

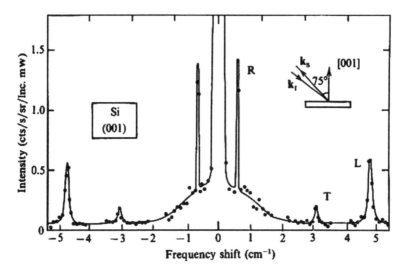

Figure 2.7. Brillouin spectrum for scattering from phonons at the planar surface of Si obtained in an oblique-incidence backscattering geometry (see inset). Peaks are indicated for the longitudinal 'L', transverse 'T' and Rayleigh 'R' modes. After Sandercock [24].

phonon peaks with increasing Sb, and their dependence on the polarizations of the incident and scattered light, motivated further studies on both the experimental and theoretical side (see [11]).

The acoustic phonons at planar surfaces had previously been observed using BLS and an example is given in figure 2.7 for the phonons at the (001) surface of a Si sample [24]. An oblique-incidence backscattering geometry was employed to excite modes with $q_{\parallel} \neq 0$. The peaks corresponding to the bulk longitudinal and transverse modes, as well as the surface Rayleigh mode, are seen on both the anti-Stokes and Stokes side of the spectrum, in accordance with the theory discussed in subsection 2.2.1. Another example of BLS from acoustic phonons is given in figure 2.8, taking the case of a spherical sample of SiO_2. The relevant theory (see subsection 2.2.2) indicates that the mode frequencies depend on integer quantum numbers n and ℓ with the latter being related to angular momentum. In figure 2.8 the acoustic-phonon peaks are labelled as $S_{n\ell}$ and $A_{n\ell}$ on the Stokes and anti-Stokes sides, respectively. These are all spheroidal modes, since the torsional modes are precluded by the selection rules of BLS. For the same reason the $\ell = 1$ spheroidal modes are not observed. Overall the measured frequencies are in good agreement with theory.

Next we show some results for phonon dispersion relations obtained by Doak *et al* by inelastic scattering of He atoms at the surface of Ag (111). The solid and open circles in figure 2.9 are for measurements on two differently prepared samples, but give essentially the same results for the phonon energies expressed here in MeV. Note that 1 MeV converts to approximately 242 GHz or 8.07 cm^{-1} in other energy- and frequency-related units. The data refer to scans across different directions in the 2D Brillouin zone (see the inset for the notation). The theory lines labelled 'X', 'Y' and 'Z' were calculated using a lattice-dynamical model for fcc crystals by Armand [27]

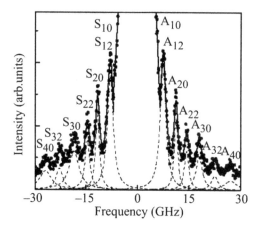

Figure 2.8. Brillouin spectrum showing scattering from acoustic phonons in SiO_2 nanospheres with diameter 340 nm. The peaks are fitted with Lorentzian functions (dashed curves) and the resultant fitted spectrum (solid line) is compared with the experimental data. See the text for labelling of the confined acoustic phonons. After Kuok *et al* [25].

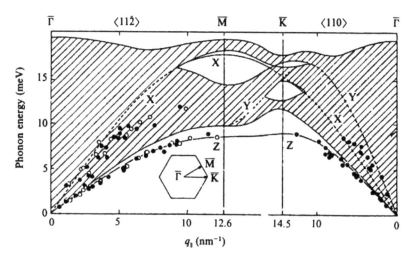

Figure 2.9. Results for phonon energy versus in-plane wave vector $q_{\parallel}$ for scattering of He atoms from the surface of Ag (111). The solid and open circles refer to solid and epitaxial crystal samples, respectively. The bulk continuum is shaded and theory lines 'X', 'Y' and 'Z' are included. After Doak *et al* [26].

and they refer to phonon modes with different polarizations. Agreement between experiment and theory is good for the lower mode ('Z') which is the extension of the Rayleigh mode across the 2D Brillouin zone. The theory is less successful for the other modes ('X' and 'Y') but improvements were subsequently made by Bortolani *et al* [28] using modified force constants.

The final example in this section will illustrate inelastic electron scattering, which was pioneered as a technique for studying surface phonons by Lehwald *et al* [29]. After the initial studies for clean surfaces the techniques were applied to planar

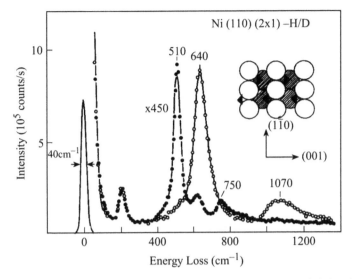

Figure 2.10. Electron energy loss scattering from phonons for hydrogen, H, (open circles) and deuterium, D, (full circles) adsorbed on Ni (110). The instrumental resolution is 40 cm^{-1}. The surface geometry is shown in the inset. After Ibach [30].

surfaces patterned with adsorbed atoms. In figure 2.10 the electron energy loss spectra for phonons are shown for a Ni (110) surface with adsorbed H and D atoms in the (2×1) arrangement depicted in the inset [30]. The main scattering peaks (at 640 cm^{-1} for H and at 510 cm^{-1} for D, as marked) are due to transitions between quantized vibrational states of the absorbed atoms. The ratio of these two frequencies is close to 1.25, which is substantially different from the value $\sqrt{2}$ that might be expected for simple classical vibrations of D and H atoms with a mass ratio of 2. The discrepancy can be explained in terms of strong quantum-mechanical effects associated with vibrations of the lighter-mass H atoms [30].

2.4 Phonons in multilayers and superlattices

In the final two sections of this chapter on phonons we consider nanostructured arrays with multiple interfaces between two or more materials. First in this section we describe theoretical and experimental studies of phonons in multilayers as depicted, for example, in figure 1.4(b). Then in section 2.5 we turn our attention to the case of lateral arrays, as in phononic crystals (see, e.g., figure 1.4(c)).

2.4.1 Theoretical results

As before in this chapter, the phonons in periodic superlattices may be studied in terms of either the lattice dynamics or elastic waves (continuum acoustics). We start here with the latter method, since it is typically simpler to apply and also we may follow an analogous approach to our treatment of optical waves in subsection 1.5.1. There we used a bulk-slab model for the system and properties of the transfer matrix for the excitations.

We may employ the same geometry for an infinite two-component...$ABABABAB$... superlattice as in figure 1.8, except that for acoustic phonons the layers will be labelled in terms of density (ρ_A or ρ_B) and elastic constant (C_A or C_B) instead of dialectic constant. Within the layers the 1D wave equation is satisfied:

$$\rho_i \frac{\partial^2 u}{\partial t^2} = C_i \frac{\partial^2 u}{\partial z^2}, \quad i = A, B. \tag{2.38}$$

Here the wave propagation is in the z-direction with velocity $v_i = (C_i/\rho_i)^{1/2}$ and the displacement u may be either longitudinal or transverse. The boundary conditions at each interface are that u and $C\partial u/\partial z$ are continuous (see sections 2.1 and 2.2). For simplicity, we are assuming no variations in the y- and z-directions.

The theory now proceeds just as in the optical-wave example in subsection 1.5.1, except that u replaces an electric-field component and the acoustic-phonon dispersion relations replace the optics result. Thus the Bloch wavenumber Q is introduced through $u(z + D) = \exp(iQD)u(z)$, replacing equation (1.46). Also the wave-equation solutions for u at the upper and lower interfaces of either an A or B layer can be written down by analogy with equations (1.47) and (1.49) where now $q_{iz} = \omega/v_i$ ($i = A, B$). Then 2×2 matrices F_i and X_i can be written down as in equations (1.52) and (1.55), and the formal results related to the transfer matrix T in (1.56)–(1.62) still hold. Specifically, the Rytov equation, analogous to (1.64), for the dispersion relation is [31]

$$\cos(QD) = \cos(q_{Az}d_A)\cos(q_{Bz}d_B) - g\sin(q_{Az}d_B)\sin(q_{Bz}d_B), \tag{2.39}$$

with

$$g = (1 + Z^2)/2Z. \tag{2.40}$$

Here Z is the ratio of acoustic impedances for the two media:

$$Z = C_A q_{Az}/C_B q_{Bz} = \sqrt{\rho_A C_A/\rho_B C_B}. \tag{2.41}$$

We note that if the media are impedance 'matched', meaning $Z = 1$, then the Rytov equation simplifies to $\cos(QD) = \cos(q_{Az}d_A + q_{Bz}d_B)$, which has the solutions

$$\omega = \pm QD + n\pi, \quad (n = \text{integer}). \tag{2.42}$$

The interpretation is broadly similar to that described in subsection 1.5.1, corresponding to a folding back of the acoustic-phonon excitation in the reduced Brillouin zone. More generally, when $Z \neq 1$ we have $[(1 + Z^2)/2Z] > 1$ in equation (2.39) and this leads to the opening up of stop bands (or band gaps) at the zone centre $Q = 0$ and zone edges $Q = \pm\pi/D$. This behaviour is illustrated in figure 2.11 where dispersion branches are shown for $Z = 2$ (solid lines) and $Z = 1$ (broken lines). In the case of $Z \neq 1$ stop bands occur, which are particularly evident in this example at the zone edge. No stop bands are seen when $Z = 1$ in accordance with equation (2.42). A consequence of the multiple branches, due to the folding back, is that several peaks may now be observed in a BLS experiment (see below).

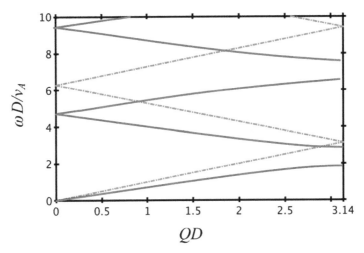

Figure 2.11. The folded acoustic-phonon dispersion curves for a two-component superlattice calculated using equation (2.39) in the elastic-wave approximation. The scaled frequency (in terms of $\omega D/v_A$ is plotted against QD across the first mini-Brillouin zone. The parameters are chosen to illustrate cases of impedance matching (broken lines) $[C_A = 2C_B,\ \rho_A = 0.5\rho_B$, implying $Z = 1$] and impedance mismatch (solid lines) $[C_A = 2C_B, \rho_A = 2\rho_B$, implying $Z = 2$]. We have taken $d_A = 2d_B = 2D/3$ in both cases.

There are two main respects in which the above type of theory has been generalized. One respect is to consider superlattice stacks that are semi-infinite or finite in the z-direction, rather than infinite. Then the occurrence of localized super-lattice modes becomes possible, in which there is a decay characteristic from one cell (of thickness D) with distance away from an external surface rather than those given by Bloch's theorem. As expected (e.g., by analogy with figure 2.4), the surface modes, if they exist, lie within the stop bands. The second type of generalization is to include modes with a wave-vector component $\mathbf{q}_{\parallel} \neq 0$ in the xy-plane. There is then a distinction between s- and p-polarizations, as discussed in simpler cases earlier. Some calculations that explore these cases can be found, for example, in [32–34]. A dispersion relation for longitudinal and p-polarized transverse phonons, including surface modes, is shown in figure 2.12, where the reduced frequencies are now plotted in terms of $q_{\parallel}D$. The properties of the surface modes are seen to depend on whether it is the Al layer or the W layer that is located at the superlattice surface.

Next we briefly discuss some of the results obtained for superlattice phonons using the lattice-dynamics approach introduced in section 2.1. Most of the interest has been in the acoustic and optic phonons in structures involving III–V semi-conductors such as GaAs. For a 1D model (applicable when the in-plane wave vector $\mathbf{q}_{\parallel} = 0$) a simple approach based on figure 2.3(b) may be utilized. Here each repeating cell consists of length d_A of material A (with alternating masses m_2 and M) and a length d_B of material B (with masses m_1 and M). This approach, which involves a direct extension of results in subsection 2.1.2 for diatomic chains combined with the transfer-matrix formalism, was used by Jusserand *et al* [43] and Djafari-Rouhani *et al* [36] to study the superlattice phonons (see also [7] for a

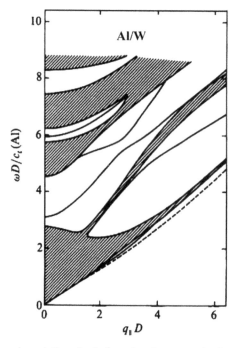

Figure 2.12. Calculated dispersion relations for bulk and surface acoustic phonons in a semi-infinite Al/W superlattice. The regions of bulk modes are shown shaded and the surface modes when an Al (W) layer is at the superlattice surface are shown by the full (broken) lines. It has been assumed that $d_{Al} = d_W = 0.5D$. After Djafari-Rouhani *et al* [33].

outline derivation). One of the main results is that equation (2.39) still applies, provided g is redefined as

$$g = \frac{1}{2}\left(\frac{\tan(q_A a)}{\tan(q_B a)} + \frac{\tan(q_B a)}{\tan(q_A a)}\right), \tag{2.43}$$

where a denotes the separation between adjacent atoms along the chain. The values of a and the force constant C are assumed, for convenience, to be the same for all nearest-neighbour atoms. In equation (2.43), as well as in the Rytov equation, q_A and q_B are related to the angular frequency ω through equation (2.9) with the appropriate masses. One possibility that arises is that q_A may be real and q_B imaginary (or vice versa) for certain values of ω. Then the superlattice modes, which are wave-like in the z-direction in one medium and decaying in the other medium, are known as *confined phonon* modes. The above theory provides a simple description of GaAs/Al$_x$Ga$_{1-x}$As superlattices, where $0 < x < 1$ for the Al concentration. Mass M would represent the As mass, while m_1 and m_2 represent the Ga and average Al$_x$Ga$_{1-x}$ masses. Extensions of the above lattice-dynamics theory to surface modes in semi-infinite superlattices and 3D models have been carried out (see, e.g., [36] and [37], respectively).

Another type of extension, carried out using both elastic continuum theory and the lattice-dynamics approach, has been to quasi-periodic superlattices, which were

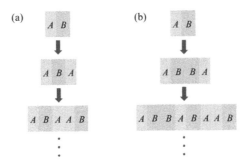

Figure 2.13. Generations of two different quasi-periodic superlattice structures: (a) Fibonacci and (b) Thue–Morse.

mentioned briefly in subsection 1.2.2. Some early accounts covering both theory and experiment were given by MacDonald [38] and Merlin [39], while a comprehensive review can be found in [40]. We will discuss mainly two-component superlattices, where there are two building blocks (thin films in the present context) labelled A and B as before. Instead of forming an alternating $ABABAB...$ as in the periodic case, the building blocks are ordered according to a growth rule such as that provided by the Fibonacci sequence mentioned earlier. By analogy with equation (1.30) we could define a growth rule recursively such that the nth stage S_n is formed according to $S_n = S_{n-1}S_{n-2}$ for $n > 1$, starting with $S_0 \equiv B$ and $S_1 \equiv A$. Hence for $n > 1$ the first few generations are $S_2 = AB$, $S_3 = ABA$, $S_4 = ABAAB$, $S_5 = ABAABABA$ and so on. The number of individual blocks in the nth generation is just the Fibonacci number F_n given by equation (1.30). In general, for each generation the number of B blocks is different from the number of A blocks: the ratio tends to the golden mean number $(1 + \sqrt{5})/2 \approx 1.618$ as $n \to \infty$. Another example of a quasi-periodic structure is generated recursively from the Thue–Morse sequence with $S_n = S_{n-1}S_{n-1}^+$ and $S_n^+ = S_{n-1}^+ S_{n-1}$ for $n > 0$, taking $S_0 = A$ and $S_0^+ = B$. The next Thue–Morse generations are $S_1 = AB$, $S_2 = ABBA$, $S_3 = ABBABAAB$ and so on, which demonstrates the property that there are equal numbers of A and B blocks in this case. The two quasi-periodic structures mentioned here are illustrated in figure 2.13.

The modification of the transfer-matrix theory described earlier for periodic superlattices to apply to various quasi-periodic superlattices is described in detail in [40]. These quasi-periodic structures have been intensively studied because of their unusual localization and the multifractal properties of the excitations.

2.4.2 Experimental results

The most successful experimental technique for studying the folded acoustic phonons and confined optical phonons in periodic superlattices has been inelastic light scattering and we mention in particular seminal work by Colvard et al [41, 42] and Jusserand et al [43]. A good review is given by Klein [44].

An example of RS from a periodic $GaAs/Al_{0.3}Ga_{0.7}As$ superlattice is shown in figure 2.14. Two spectra are plotted corresponding to (x, x) and (x, y) for the electric-field polarizations of the incident and scattered light beams. The selection

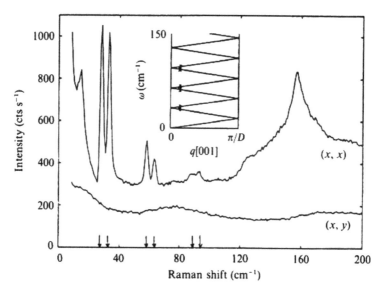

Figure 2.14. Spectra for RS from a periodic superlattice with alternating GaAs (4.2 nm) and $Al_{0.3}Ga_{0.7}As$ (0.8 nm) layers. A backscattering geometry for two different polarizations was employed. The inset shows the folded dispersion curves calculated for LA phonons, with crosses at the peak positions for (x, x) polarization. After Colvard *et al* [42].

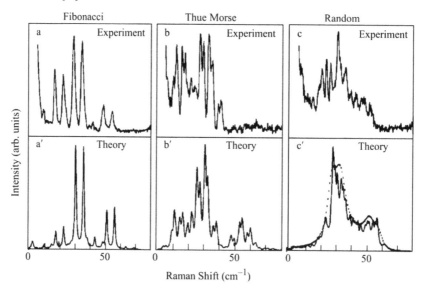

Figure 2.15. Experimental (upper) and theoretical (lower) spectra for RS from acoustic phonons in GaAs/AlAs sublattices of different structures: Fibonacci quasi-periodic (left), Thue–Morse quasi-periodic (centre) and random (right). After Merlin *et al* [46].

rule, deduced from symmetry arguments, is that RS from phonons should occur only in 'diagonal' polarizations [45] and this is clearly verified here where the phonon peaks are seen to occur in the (x, x) plot only. The multiple TA phonon peaks, having Raman shifts between about 25 and $100\,cm^{-1}$, are related to the mode folding in the inset. Next we illustrate in figure 2.15 some experimental data and theory for two different quasi-periodic GaAs/AlAs superlattices (Fibonacci and Thue–Morse), along with a comparison to a two-component superlattice where the GaAs and AlAs building blocks are chosen randomly. It is evident that the spectra for two quasi-periodic structures are different from one another and quite distinct in features from the random superlattice.

2.5 Phononic crystals

As explained earlier, phononic crystals are formed from *lateral* arrays of elements, typically periodic with a repeating cell of order several nano- or micrometres (see figure 1.4(c), for example). From the physics perspective, much the same effect can be achieved through a periodic patterning of a planar surface instead of an alternation of two different materials. A prototype phononic crystal was made by Dutcher *et al* [47] by forming a holographic grating of grooves with separation $D = 0.25\ \mu$m on a Si (001) surface. The system has 1D translational symmetry in the direction parallel to the grooves and the mini-Brillouin-zone boundaries associated with the in-plane wave vector perpendicular to the grooves are at $m\pi/D$, where m is an integer. By analogy with the discussion in the previous section for superlattices, we can describe the phonon dispersion in a reduced-zone scheme in terms of folded modes with band gaps opening at the grating zone boundaries. Some data for the frequency of the Rayleigh surface mode (introduced for a planar surface in subsection 2.2.1) plotted versus the in-plane wave vector are shown in figure 2.16. They clearly demonstrate the mode folding and provide good consistency with an earlier theory for the elastic modes at surfaces with 1D gratings [48].

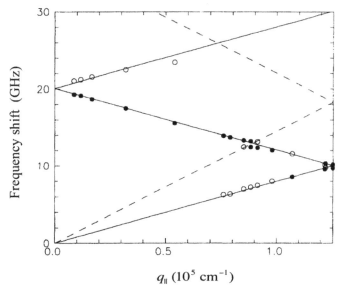

Figure 2.16. Frequency shift versus in-plane wave vector $q_\parallel$ in a reduced-zone scheme for BLS from the surface of a Si (001) surface with grating grooves. The experimental points are the open and closed circles. The solid lines correspond to the folded surface Rayleigh wave and (for comparison) the dashed lines refer to the LA wave for an unpatterned Si (100) surface. After Dutcher *et al* [47].

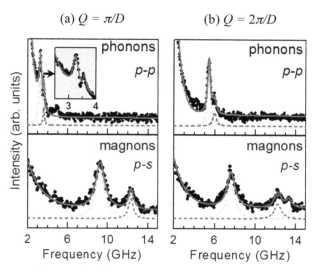

Figure 2.17. BLS spectra for scattering from phonons and magnons in different polarizations in a dual phononic/MC composed of alternating Fe and $Ni_{80}Fe_{20}$ (permalloy) nanostripes at (a) $Q = \pi/D$ and (b) $Q = 2\pi/D$ for the in-plane Bloch wave vector. After Zhang *et al* [50].

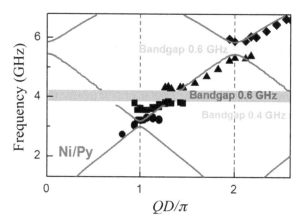

Figure 2.18. Plot of frequency versus reduced wave vector QD/π for acoustic phonons as deduced from BLS data (see figure 2.17) for a dual phononic/MC composed of alternating Fe and $Ni_{80}Fe_{20}$ (permalloy) nanostripes. After Zhang *et al* [50].

In the last decade there has been an intense interest in the bicomponent phononic crystals formed from lateral arrays (with either 1D or 2D periodicity) of two materials. For example, Gorishnyy *et al* [49] clearly demonstrated the effects of artificial band gaps opening up in a 2D hypersonic phononic crystal fabricated from a triangular-lattice arrangement of cylindrical holes in a 6 μm thick epoxy film. The holes were separated (centre-to-centre distance) by 1.36 μm and different hole radii were used to vary the sample porosity from close to zero up to about 40%. More recently a particularly thorough investigation was carried out by Zhang *et al* [50] using samples in the specific alternating-stripe geometry of figure 1.4(c). For the materials they employed Fe and $Ni_{80}Fe_{20}$ (permalloy), which allowed them to study using BLS in different polarizations both the phonon and magnon properties in the artificial crystal. Some of their data are shown in figures 2.17 and 2.18 for a sample with periodicity length $D = 500$ nm. In figure 2.17 the BLS spectra are displayed for two values π/D and $2\pi/D$ of the 1D Bloch wave vector Q (corresponding to the edges of the first and second Brillouin zone, respectively). The acoustic-phonon results, which are in the upper panels of the figure, are seen in diagonal polarization (denoted here by 'p-p'). Then the frequency versus Q plots for the phonon peaks are shown in figure 2.18, where there are band gaps at $QD/\pi = 0$, 1 and 2, as expected. The full lines (in blue) are theory using numerical simulations with a finite-element approach. The magnon results will be discussed in the next chapter. The same group subsequently generalized their phonon studies to bicomponent structures with a chess (or checker) board array, i.e., a 2D phononic crystal [51].

References

[1] Kittel C 2004 *Introduction to Solid State Physics* 8th edn (New York: Wiley)
[2] Ashcroft N W and Mermin N D 1976 *Solid State Physics* (New York: Saunders College)
[3] Srivastava G P 1990 *The Physics of Phonons* (New York: Taylor and Francis)
[4] Born M and Huang K 1998 *Dynamical Theory of Crystal Lattices* (Oxford: Oxford University Press)
[5] Maradudin A A 1981 *Festkörperprobleme* vol 21 ed J Treusch (Braunschweig: Vieweg) p 25
[6] Stegeman G I and Nizzoli F 1984 *Surface Excitations* ed V M Agranovich and R Loudon (Amsterdam: North-Holland) p 195
[7] Cottam M G and Tilley D R 2005 *Introduction to Surface and Superlattice Excitations* 2nd edn (Bristol: IOP)
[8] Wallis R D 1957 *Phys. Rev.* **105** 540
[9] Mazur P and Maradudin A A 1981 *Phys. Rev.* B **24** 2996
[10] Cottam M G and Maradudin A A 1984 *Surface Excitations* ed V M Agranovich and R Loudon (Amsterdam: North-Holland) p 1
[11] Esser N and Richter W 2000 *Light Scattering in Solids* vol 8 ed M Cardona and G Güntherodt (Berlin: Springer)
[12] Ataca C, Topsakal M, Aktürk E and Ciraci S 2011 *J. Phys. Chem.* **115** 16354
[13] Landau L D and Lifshitz E M 1970 *Theory of Elasticity* (Oxford: Pergamon)
[14] Albuquerque E L 1980 *J. Phys. C: Solid State Phys.* **13** 2623
[15] Albuquerque E L, Loudon R and Tilley D R 1980 *J. Phys. C: Solid State Phys.* **13** 1775
[16] Maradudin A A, Wallis R F, Mills D L and Ballard R L 1972 *Phys. Rev.* B **6** 1106
[17] Lamb H 1882 *Proc. Lond. Math. Soc.* **13** 189
[18] Nishiguchi N and Sakuma T 1981 *Solid State Commun.* **38** 1073
[19] Mathews J and Walker R L 1969 *Mathematical Methods of Physics* (New York: Addison-Wesley)
[20] Hudson G E 1943 *Phys. Rev.* **63** 46
[21] Viktorov I A 1958 *Sov. Phys. Acoust.* **4** 131
[22] Rulf B 1969 *J. Acoust. Soc. Am.* **45** 493
[23] Hünermann M, Geurts J and Richter W 1991 *Phys. Rev. Lett.* **66** 640
[24] Sandercock J R 1978 *Solid State Commun.* **26** 547
[25] Kuok M H, Lim H S, Ng S C, Liu N N and Wang Z K 2003 *Phys. Rev. Lett.* **90** 255502

[26] Doak R B, Harten U and Tonnies J P 1983 *Phys. Rev. Lett.* **51** 578

[27] Armand G 1983 *Solid State Commun.* **48** 261

[28] Bortolani V, Franchini A, Nizzoli F and Santoro G 1984 *Phys. Rev. Lett.* **52** 429

[29] Lehwald S, Szeftel J M, Ibach H, Rahman T S and Mills D L 1983 *Phys. Rev. Lett.* **50** 518

[30] Ibach H 1991 *Phys. Scripta* **T39** 323

[31] Rytov S M 1956 *Sov. Phys. Acoust.* **2** 68

[32] Camley R E, Djafari-Rouhani B, Dobrzynski L and Maradudin A A 1983 *Phys. Rev.* B **27** 7318

[33] Djafari-Rouhani B, Dobrzynski L, Hardouin-Duparc O, Camley R E and Maradudin A A 1983 *Phys. Rev.* B **28** 1711

[34] Babiker M, Tilley D R, Albuquerque E L and Gonçalves da Silva C E T 1983 *J. Phys. C: Solid State Phys.* **18** 1269

[35] Jusserand B, Paquet D, Kervarec J and Regreny A 1984 *J. Phys.* **45** C5–154

[36] Djafari-Rouhani B, Sapriel J and Bonnouvrier F 1985 *Superlattices Microstruct.* **1** 29

[37] Molinari E, Baroni S, Giannozzi P and de Gironcoli S 1992 *Phys. Rev.* B **45** 4280

[38] MacDonald A H 1988 *Interfaces, Quantum Wells, and Superlattices* ed C R Leavens and R Taylor (New York: Plenum) p 347

[39] Merlin R 1989 *Light Scattering in Solids* vol 5 ed M Cardona and G Güntherodt (Berlin: Springer) p 214

[40] Albuquerque E L and Cottam M G 2003 *Phys. Rep.* **376** 225

[41] Colvard C, Merlin R, Klein M V and Gossard A C 1980 *Phys. Rev. Lett.* **45** 298

[42] Colvard C, Gant T A, Klein M V, Merlin R, Fischer R, Morkoc H and Gossard A C 1985 *Phys. Rev.* B **31** 2080

[43] Jusserand B, Paquet D and Regreny A 1984 *Superlattices Microstruct.* **1** 61

[44] Klein M V 1986 *IEEE J. Quantum Elect.* **QE-22** 1760

[45] Hayes W and Loudon R 1978 *Scattering of Light by Crystals* (New York: Wiley)

[46] Merlin R, Bajema K, Nagle J and Ploog K 1987 *J. Phys.* **48** C5–503

[47] Dutcher J R, Lee S, Hillebrands B, McLaughlin G J, Nickel B G and Stegeman G I 1992 *Phys. Rev. Lett.* **68** 2464

[48] Glass N E and Maradudin A A 1983 *J. Appl. Phys.* **54** 796

[49] Gorishnyy T, Ullal C K, Maldovan M, Fytas G and Thomas E L 2005 *Phys. Rev. Lett.* **94** 115501

[50] Zhang V L, Ma F S, Pan H H, Lin C S, Lim H S, Ng S C, Kuok M H, Jain S and Adeyeye A O 2012 *Appl. Phys. Lett.* **100** 163118

[51] Zhang V L, Hou C G, Pan H H, Ma F S, Kuok M H, Lim H S, Ng S C, Cottam M G, Jamali M and Yang H 2012 *Appl. Phys. Lett.* **101** 053102

Chapter 3

Magnons

In this chapter we consider magnons, which are the quanta of SWs associated with the fluctuating magnetization in ordered magnetic materials at temperatures normally well below their critical temperature. Just as in the phonon case, there are different approaches (and different regimes of behaviour). An important consideration is the relative importance of the dynamical effects due to the quantum-mechanical exchange interactions and the electromagnetic magnetic dipole–dipole interactions, among other factors, and this depends in part on the wave vector of the excitations.

Some general references on magnons (or SWs, since we will mostly use the two terms interchangeably as is common in the literature) are [1–3]. Some additional references where the surface and/or interface properties of magnons are emphasized include [4–8]. Here we start with an overview of the interactions that are important for the static and dynamical behaviour of ordered magnetic materials, before proceeding with studies of the magnetic excitations.

3.1 Regimes of magnetization dynamics

The ordered magnetic materials of interest in this chapter are ferromagnets, ferrimagnets and antiferromagnets, whose basic characteristics are described in textbooks on solid-state physics (see, e.g., [9–11]). Typically ferromagnets consist of a lattice of identical magnetic moments or spins that are aligned parallel to one another at low temperatures. In contrast, in ferrimagnets and antiferromagnets there are two (or more) sublattices with some of the magnetic moments being aligned antiparallel. In a two-sublattice antiferromagnet the magnetic moments on each sublattice are equal in magnitude but opposite in direction (in the case of no externally applied field) and so the net magnetization is zero. In a ferrimagnet the magnetic moments of the sublattices do not cancel and hence there is a nonzero magnetization. The critical temperature, above which the long-range magnetic order essentially disappears, is known as the Curie temperature T_C in ferromagnets and ferrimagnets and the Néel temperature T_N in antiferromagnets.

It is well known that the dominant magnetic interaction, which is responsible for the overall static magnetic order in these materials, is the quantum-mechanical *exchange interaction* (see, e.g., [9, 10]). In the simplest case this arises from the overlap of the quantum-mechanical wave functions of two electrons on neighbouring atoms. The spin state for any electron pair may be either symmetric (spins parallel) or antisymmetric (spins antiparallel) with respect to an interchange. From the Pauli exclusion principle it follows that the corresponding spatial part of the wave function must be antisymmetric or symmetric, respectively, to ensure the antisymmetry of the total wave function. As a consequence of the Coulomb interaction between the electrons it can be shown that the expectation value of the energy for the electron pair (e.g., as calculated in Heitler–London theory) is different in the two states, with the difference usually expressed in the form $-J_{12}\mathbf{S}_1 \cdot \mathbf{S}_2$. Here $\mathbf{S}_1$ and $\mathbf{S}_2$ are the spin operators for the two electrons while J_{12}, which depends on the overlap of the electronic wave functions, is known as the exchange integral. Depending on the sign of the exchange, we may have either ferromagnetism ($J_{12} > 0$) or antiferromagnetism ($J_{12} < 0$). The arguments may be extended to other multi-electron situations and to ferrimagnetism.

The concept of SWs in bulk exchange-dominated ferromagnets was introduced by Bloch [12] using a semi-classical approach with spin vectors. He envisioned an excitation with one spin reversal being 'shared' over many spins throughout a crystal in terms of a collective wave-like disturbance with precessing spins. The amplitude of spin precession would be constant for all the equivalent spin sites in a bulk material, but the phase angles would be different. A wide range of quantum and semi-classical SW approaches are now employed, as we discuss in the following sections.

By contrast with the short-range exchange, we will also in general have energy contributions from the long-range magnetic dipole interactions proportional to

$$g^2 \mu_B^2 \left(\frac{\mathbf{S}_1 \cdot \mathbf{S}_2}{r_{12}^3} - \frac{3(\mathbf{S}_1 \cdot \mathbf{r}_{12})(\mathbf{S}_2 \cdot \mathbf{r}_{12})}{r_{12}^5} \right) \tag{3.1}$$

between spins 1 and 2 separated by vector $\mathbf{r}_{12}$. Here g and μ_B denote the Landé factor and Bohr magneton, respectively. Although the individual dipole–dipole interaction strengths may typically be much smaller than the nearest-neighbour exchange energy, the dipole–dipole effects will become important (and eventually will be dominant) at sufficiently small excitation wave vectors. This is because of their long-range character. As we will show later in this chapter, the exchange contribution to the energy of a ferromagnetic excitation with wave vector $\mathbf{q}$ is of order $SJ\,|\mathbf{q}|^2 a^2$ for $|\mathbf{q}|a \ll 1$ (with J denoting the nearest-neighbour exchange and S the spin quantum number) while the dipolar contribution to the energy is of order $g\mu_B \mu_0 M_0$ (with M_0 denoting the static magnetization). Hence the two terms are comparable for

$$|\mathbf{q}|a \sim \sqrt{g\mu_B \mu_0 M_0/SJ} \ll 1. \tag{3.2}$$

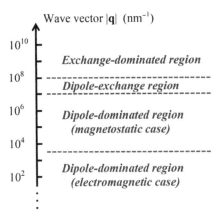

Figure 3.1. Schematic representation of the different regions of SW behaviour for a ferromagnet in terms of its wave vector. Note that the value 10^{10} nm^{-1} corresponds roughly to a Brillouin-zone-boundary wave vector.

The dipolar terms dominate below this magnitude of wave vector and the exchange terms dominate above. There is a 'crossover region' where both terms need to be included and this is known as the *dipole-exchange region* as indicated in figure 3.1.

Since the magnetic dipole–dipole interactions are important only at long wavelengths (compared to the lattice parameter a) they are often treated in a continuous-medium representation using the magnetic-field variables of electromagnetism together with Maxwell's equations, rather than the energy representation as in equation (3.1). We will use both approaches in this book, in particular for discussion of the dipole-exchange region, and we illustrate the connection between them. In the dipole-dominated region two cases arise as indicated. They correspond to either using a magnetostatic approximation for Maxwell's equations at larger wave vectors in this range or employing the electromagnetic equations with retardation effects included as required at very small wave vectors.

In this chapter we will describe the exchange-dominated, dipole-exchange and magnetostatic regions. The electromagnetic region corresponds to a situation where the magnons are strongly coupled to the photons to give mixed excitations known as magnon-polaritons. They will be treated along with other types of polaritons in chapter 5.

3.2 Exchange-dominated waves in films

3.2.1 Ferromagnets

For simplicity we start by considering the magnons or SWs in ferromagnetic films and for the present we ignore the dipole–dipole interactions. Other types of magnetic behaviour and geometries will be introduced later. The Heisenberg spin Hamiltonian for the system may be expressed as

$$H = -\frac{1}{2}\sum_{i,j} J_{ij}\mathbf{S}_i \cdot \mathbf{S}_j - g\mu_B B_0 \sum_i S_i^z, \qquad (3.3)$$

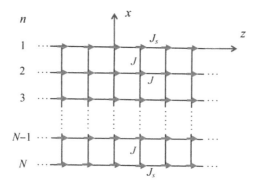

Figure 3.2. Geometry and coordinate axes for a Heisenberg ferromagnetic film. Integer n labels the atomic layers (in the yz-plane) parallel to the surfaces, while J and J_s are the bulk and surface nearest-neighbour exchange constants.

where i and j label all the spin sites, J_{ij} is the corresponding exchange interaction and the factor of $1/2$ before the first term allows for double counting in the summation. The second term represents the Zeeman energy of the spins in an applied magnetic field of magnitude B_0 in a fixed direction (taken as the z-axis). The film geometry is illustrated in figure 3.2, where we have assumed an sc lattice structure with lattice parameter a. The parallel surfaces of the film are in the yz-plane and the N atomic layers are labelled by integer n ($= 1, 2, 3, \ldots, N$) as shown. The film is infinite in the y- and z-directions. We adopt a simple model where the exchange couples nearest neighbours only, having the value J_s between two spins in a surface layer and the bulk value J otherwise. The equilibrium orientation of the spins is assumed to be in-plane and along the z-direction of the applied magnetic field.

There are various ways to proceed in calculating the magnons. Here we will work directly in terms of spin operators and their quantum-mechanical equation of motion. Other techniques, such as those based on the torque equation of motion for the magnetization vector or a transformation to boson operators, will be utilized later. First we introduce the usual spin-raising S_j^+ and spin-lowering S_j^- operators at any site j by

$$S_j^\pm = S_j^x \pm iS_j^y. \tag{3.4}$$

The commutation relations satisfied by these operators, together with the other component S_j^z can then be expressed as (see, e.g., [13])

$$\left[S_i^+, S_j^-\right] = 2S_j^z \delta_{ij}, \qquad \left[S_i^z, S_j^\pm\right] = \pm S_i^\pm \delta_{ij}, \tag{3.5}$$

where we denote $[A, B] = AB - BA$ for any operators A and B. Also, for convenience, we employ units such that $\hbar = 1$. The equation of motion [13] for the time dependence of any operator A was given in equation (1.36) in terms of its commutator with H.

To study the magnons, which are excitations associated with the transverse spin components, it is natural that we choose to take $A = S_j^+$. Then using equations (3.3) and (3.5) we eventually obtain

$$i\frac{d}{dt}S_j^+ = g\mu_B B_0 S_j^+ + \sum_i J_{ij}\left(S_i^z S_j^+ - S_j^z S_i^+\right). \tag{3.6}$$

This is an NL differential equation in terms of the spin components, but it can be simplified (or linearized) by making the *random-phase approximation* (RPA) to replace the factor $S_i^z S_j^+ - S_j^z S_i^+$ by $\langle S_i^z\rangle S_j^+ - \langle S_j^z\rangle S_i^+$, where $\langle\cdots\rangle$ denotes a thermal average. This approximation involves 'decoupling' a product of operators by replacing each S^z operator by its scalar thermal average. The justification is that at low temperatures in an ordered magnetic material the fluctuations in the z-components will be small compared to those in the xy-plane. In fact, if we make the low-temperature replacement $\langle S_i^z\rangle \rightarrow S$ in equation (3.6) we find

$$i\frac{d}{dt}S_j^+ = g\mu_B B_0 S_j^+ + S\sum_i J_{ij}\left(S_j^+ - S_i^+\right). \tag{3.7}$$

By analogy with the procedure used for phonons in chapter 2 we seek wave-like solutions of the form

$$S_j^+ = s_n \exp\left[i\left(\mathbf{q}_\parallel \cdot \mathbf{r}_{\parallel,j} - \omega t\right)\right], \tag{3.8}$$

for a site j in layer n. The 2D vectors parallel to the film surfaces are $\mathbf{q}_\parallel = (q_y, q_z)$ and $\mathbf{r}_\parallel = (y, z)$. Equation (3.7) can be rewritten as a finite set of coupled equations in terms of n:

$$\left\{\omega - g\mu_B B_0 - SJ - 4SJ_s\left[1 - \gamma\left(\mathbf{q}_\parallel\right)\right]\right\}s_1 + SJs_2 = 0,$$

$$\left\{\omega - g\mu_B B_0 - 2SJ - 4SJ\left[1 - \gamma\left(\mathbf{q}_\parallel\right)\right]\right\}s_n + SJ(s_{n-1} + s_{n+1}) = 0 \quad (1 < n < N), \tag{3.9}$$

$$\left\{\omega - g\mu_B B_0 - SJ - 4SJ_s\left[1 - \gamma\left(\mathbf{q}_\parallel\right)\right]\right\}s_N + SJs_{N-1} = 0,$$

where $\gamma(\mathbf{q}_\parallel)$ is a structure factor for the sc lattice defined by

$$\gamma\left(\mathbf{q}_\parallel\right) = \frac{1}{2}\left[\cos\left(q_y a\right) + \cos\left(q_z a\right)\right]. \tag{3.10}$$

The above equations are similar to those encountered in the lattice-dynamics calculations in chapter 2 and so the same methods can be employed. These include seeking solutions for $s_n(\mathbf{q}_\parallel)$ formed by the superposition of two wave-like terms (one incident at a surface and one reflected) on the one hand or using the TDM method (see section A.1 of the appendix) on the other. Following the former approach we write for the x-dependence

$$s_n = A\exp\left[-iq_x(n-1)a\right] + B\exp\left[iq_x(n-1)a\right], \tag{3.11}$$

where we note that the x-coordinate for layer n of the film is $(1 - n)a$. It is now straightforward to verify by substitution that equation (3.11) is a solution of (3.9) provided $\omega = \omega_B(\mathbf{q})$ where $\mathbf{q} = (q_x, \mathbf{q}_{\parallel})$ and the bulk-magnon dispersion relation is found as

$$\omega_B(\mathbf{q}) = g\mu_B B_0 + 2SJ\Big[3 - \cos\big(q_x a\big) - \cos\big(q_y a\big) - \cos\big(q_z a\big)\Big]. \tag{3.12}$$

The first and last of the coupled equations in (3.9), i.e., those for $n = 1$ and $n = N$, constitute the boundary conditions. Each can be used to find an expression for the ratio B/A and the consistency condition between them gives

$$\tan(q_x L) = \frac{i\big[\Delta - \exp(iq_x a)\big]^2 - i\big[\Delta - \exp(-iq_x a)\big]^2}{\big[\Delta - \exp(iq_x a)\big]^2 + \big[\Delta - \exp(-iq_x a)\big]^2}, \tag{3.13}$$

where the film thickness is $L = (N - 1)a$ and Δ is a surface-dependent parameter:

$$\Delta = \left(1 + \frac{4(J - J_s)}{J}\Big[1 - \gamma\big(\mathbf{q}_{\parallel}\big)\Big]\right)^{-1}. \tag{3.14}$$

Equation (3.13) is an implicit equation for q_x which can only be satisfied by a set of discrete values. Hence it can be interpreted as a quantization condition both for the wave-vector component q_x and consequently for the bulk-magnon frequency ω_B.

Similar arguments may be used to study surface magnons by seeking localized (or attenuated) solutions for s_n of the form

$$s_n = C_1 \exp(-\kappa n a) + C_2 \exp(\kappa n a), \tag{3.15}$$

where $\mathrm{Re}(\kappa) > 0$. For a thick film (in the semi-infinite limit where $N \rightarrow \infty$) it can easily be shown that the surface magnon frequency consists of a single branch $\omega = \omega_S(\mathbf{q}_{\parallel})$ with

$$\omega_S\big(\mathbf{q}_{\parallel}\big) = g\mu_B B_0 + 4SJ\Big[1 - \gamma\big(\mathbf{q}_{\parallel}\big)\Big] - SJ\frac{(1 - \Delta)^2}{\Delta}. \tag{3.16}$$

This solution only exists provided $|\Delta| < 1$ where Δ is defined in equation (3.14). In fact the attenuation factor κ for the mode is found from $\exp(-\kappa a) = |\Delta|$ in this case. There are two ways in which the existence condition can be satisfied. One occurs when $J_s < J$, making $0 < \Delta < 1$ for all $\mathbf{q}_{\parallel}$, while the other occurs when $J_s > 1.25J$, making $-1 < \Delta < 0$ for at least some values of $\mathbf{q}_{\parallel}$. These are loosely referred to as acoustic and optic surface magnons, respectively, since they have frequencies below and above the bulk-magnon region. Some numerical examples are shown in figure 3.3 (see curves A and B). Curve C illustrates the effect of adding a surface anisotropy, acting as an additional field B_{AS} on the surface layer spins only, in splitting off an acoustic surface mode from the bulk-mode continuum.

The generalization of the above surface-magnon result to Heisenberg ferromagnetic films with finite thickness (the case of N finite) is given in [14]. For most values of the parameters each surface-magnon branch is split into two and the existence

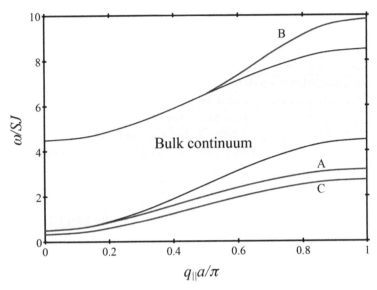

Figure 3.3. Plot of magnon frequency (in terms of ω/SJ) against $|\mathbf{q}_\| a|$ for a semi-infinite Heisenberg ferromagnet with sc lattice structure, taking $\mathbf{q}_\| = (q_\|, 0)$. The bulk-magnon continuum is shown (shaded region) and there are three surface magnon branches: A, $J_s/J = 0.5$ and $B_{AS} = 0$; B, $J_s/J = 2$ and $B_{AS} = 0$; and C, $J_s/J = 0.5$ and $g\mu_B B_{AS}/SJ = -0.5$. Also $g\mu_B B_0/SJ = 0.5$.

conditions (for localization) are modified. Typically, the mode splitting is smaller at larger wave vectors, as well as at larger N (see numerical examples given in [15]).

To conclude this topic we remark that, in a long-wavelength limit where both $|\mathbf{q}a| \ll 1$ and $|\mathbf{q}_\| a| \ll 1$, the equivalent results for the dispersion relations may be derived using a continuum (or macroscopic) method in which the finite-difference terms are replaced by a continuous magnetic variable and its derivatives. Within such an approach we conventionally employ the torque equation of motion for the magnetization as

$$\frac{d\mathbf{M}}{dt} = \gamma(\mathbf{M} \times \mathbf{B}) = \gamma\mu_0(\mathbf{M} \times \mathbf{H}) \tag{3.17}$$

when the damping is ignored. This replaces the quantum-mechanical operator equation of motion (1.36) used earlier. In the above γ is the gyromagnetic ratio and $\mathbf{M}$ is the total magnetization, which can be written as

$$\mathbf{M} = M_0\hat{\mathbf{z}} + \mathbf{m}(\mathbf{r})\exp(-i\omega t), \tag{3.18}$$

where M_0 is the static magnetization in the z-direction (unit vector $\hat{\mathbf{z}}$) and $\mathbf{m}(\mathbf{r})$ is the fluctuating part of the magnetization. We are considering the terms at angular frequency ω, so the total field has the form

$$\mu_0\mathbf{H} = B_0\hat{\mathbf{z}} + \mu_0\mathbf{h}(\mathbf{r})\exp(-i\omega t). \tag{3.19}$$

When equations (3.18) and (3.19) are substituted into (3.17) and a linearization approximation of neglecting second-order small quantities is made, we obtain

$$-i\omega\mathbf{m}(\mathbf{r}) = \gamma\hat{\mathbf{z}} \times \left[\mu_0 M_0 \mathbf{h}(\mathbf{r}) - B_0 \mathbf{m}(\mathbf{r})\right]. \tag{3.20}$$

For us to proceed we need an expression for $\mathbf{h}(\mathbf{r})$. In the present case of the exchange-dominated regime this may be deduced from the Heisenberg Hamiltonian (3.3), although we will later see that there can be additional dipolar contributions in general. The procedure adopted here is that the components of the total effective magnetic field at any site j can first be found by taking the functional derivative

$$B_j^\alpha = -\frac{1}{g\mu_B}\frac{\delta H}{\delta S_j^\alpha} \quad (\alpha = x, y, z). \tag{3.21}$$

Then, on going over to continuous variables (by spatially averaging on the microscopic scale) and noting that $g\mu_B$ can be equated with γ since we are using units such that $\hbar = 1$, it is found that (see, e.g., [1–3])

$$\mu_0 \mathbf{h}(\mathbf{r}) = \left(a^2 J/\gamma\right)\nabla^2\mathbf{m}(\mathbf{r}) \tag{3.22}$$

when $\mathbf{r}$ is in the interior of a magnetic material. However, when $\mathbf{r}$ is at a free surface (corresponding, say, to the plane $x = 0$), the result to lowest order in the spatial derivatives becomes

$$\mu_0 \mathbf{h}(\mathbf{r}) = (aJ/\gamma)\frac{\partial\mathbf{m}(\mathbf{r})}{\partial x} \quad (x = 0). \tag{3.23}$$

When equation (3.22) is substituted into (3.20) we arrive at the result that $\mathbf{m}(\mathbf{r})$ in the interior of a Heisenberg ferromagnet satisfies

$$i\omega\mathbf{m}(\mathbf{r}) = \hat{\mathbf{z}} \times \left[\gamma B_0 - D\nabla^2\right]\mathbf{m}(\mathbf{r}), \tag{3.24}$$

where we used the relationships $D = SJa^2$ and $M_0 = S/a^3$ for an sc ferromagnet at low temperature $T \ll T_C$. Equation (3.24) has wave-like solutions of the form $[A\exp(iq_x x) + B\exp(-iq_x x)]\exp(i\mathbf{q}_\parallel \cdot \mathbf{r}_\parallel)$ yielding the bulk-magnon frequency as

$$\omega_B(\mathbf{q}) = g\mu_B B_0 + D\left[q_x^2 + q_\parallel^2\right], \tag{3.25}$$

which is just the small wave-vector form of equation (3.12). The corresponding result for the surface-magnon frequency can likewise be recovered. Equation (3.25) holds for other cubic lattices (bcc and fcc) provided D is redefined.

3.2.2 Antiferromagnets

As mentioned earlier, antiferromagnets generally have two interpenetrating sublattices with equal and opposite magnetic moments, so they cancel out giving no net magnetization in zero applied magnetic field. The Heisenberg Hamiltonian can be written as

$$H = \sum_{i,j}J_{ij}\mathbf{S}_i \cdot \mathbf{S}_j - g\mu_B B_0\left[\sum_i S_i^z + \sum_j S_j^z\right] - g\mu_B B_A \sum_i S_i^z + g\mu_B B_A \sum_j S_j^z, \tag{3.26}$$

where i and j label the sites on sublattice 1 (spin-up) and 2 (spin-down), respectively, and $J_{ij} > 0$ denotes the nearest-neighbour exchange interaction between sites on opposite sublattices. The second term represents the Zeeman energy due to an applied field B_0 along the z-direction, while the last two terms account for the anisotropy energy in terms of an effective anisotropy field B_A that has the same magnitude but opposite sign on the two sublattices. Anisotropy is known to play an important role in stabilizing the antiferromagnetic order and typically arises due to crystal-field effects (see, e.g., [1]).

It is well known from standard textbooks on solid-state physics [9, 10] that there are two magnon branches in infinite antiferromagnets, essentially because there are two spin sites per unit cell, so the situation is broadly analogous to the calculations for the lattice dynamics of diatomic chains (see subsection 2.1.2) where two excitation branches were found. In the antiferromagnetic case the two bulk-magnon branches are degenerate in magnitude in zero applied magnetic field.

If a planar surface is introduced to form a semi-infinite structure the effects are more subtle for antiferromagnets than was the case for ferromagnets. This is because the surface may, at least in some cases, remove the symmetry between the up and down sublattices leading to important consequences for any localized surface excitations. To illustrate this point we will compare two lattice structures, namely, body-centred tetragonal (bct, as in the rutile-structure antiferromagnets MnF_2 and FeF_2) and sc (as in the perovskite structure antiferromagnets $RbMnF_3$ and $KNiF_3$), taking (100) surfaces in both cases. These are shown schematically in figure 3.4. It is

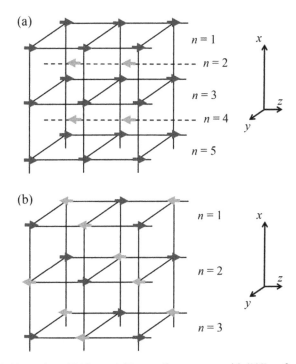

Figure 3.4. Schematic illustration of (a) bct and (b) sc antiferromagnets with (100) surfaces. The numbering of the atomic layers is shown.

evident that in case (a) each layer parallel to the surface contains magnetic sites belonging to one sublattice only in an alternating fashion. Moreover the nearest-neighbour intersublattice exchange is to sites on adjacent layers only. In contrast, in case (b) there are equal numbers of sites from both sublattices in all layers and the exchange coupling is to sites in the same as well as adjacent layers.

We begin with the bulk and surface magnons in a semi-infinite bct antiferromagnet, as in figure 3.4(a). The surface layer $n = 1$ and all other odd layers have only spin-up sites (i.e., their spin projections are along the z-direction) and the even layers have only spin-down sites (spin projections along the $-z$-direction). We may now form a set of coupled-finite difference equations for the spin operators, as was done in the ferromagnetic case, except that we employ equation (3.26) for the Hamiltonian and we take the sublattice structure into account. The results are

$$\left(\omega - g\mu_B B_A - 4SJ\right)s_1 - 4SJ\gamma_0\!\left(\mathbf{q}_{\parallel}\right)s_2 = 0,$$

$$\left(\omega + g\mu_B B_A + 8SJ\right)s_{2m} + 4SJ\gamma_0\!\left(\mathbf{q}_{\parallel}\right)(s_{2m-1} + s_{2m+1}) = 0 \quad (m \geqslant 1), \qquad (3.27)$$

$$\left(\omega - g\mu_B B_A - 8SJ\right)s_{2m+1} - 4SJ\gamma_0\!\left(\mathbf{q}_{\parallel}\right)(s_{2m} + s_{2m+2}) = 0 \quad (m \geqslant 1),$$

where J is the nearest-neighbour exchange interaction and the spin amplitudes $s_n(\mathbf{q}_{\parallel})$ are defined as in equation (3.8). Also we denote

$$\gamma_0\!\left(\mathbf{q}_{\parallel}\right) = \cos\!\left(q_y a/2\right)\cos\!\left(q_z c/2\right), \qquad (3.28)$$

where c is the lattice parameter corresponding to the tetragonal (z-)axis in the bct lattice. For simplicity we have assumed here that the applied magnetic field B_0 is zero and that the spin averages $\langle S^z \rangle$ may be approximated by the values S and $-S$ for sites on sublattices 1 and 2, respectively. The analysis is made easier if one notes that the odd-layer equations can be used to eliminate all of the odd-layer amplitudes from the remaining equations, so that a new set of finite-difference equations is obtained that involves only the even-layer amplitudes.

The solutions for the bulk and surface magnons can be investigated in the same manner as for the ferromagnetic case (see, e.g., [4, 5]), so we quote only the final results. Corresponding to the bulk magnons, wave-like solutions are found with $\omega = \pm\omega_B(\mathbf{q})$ where

$$\omega_B(\mathbf{q}) = \left[\left(g\mu_B B_A + 8SJ\right)^2 - \left\{8SJ\cos\!\left(q_x a/2\right)\gamma_0(\mathbf{q}_{\parallel})\right\}^2\right]^{1/2}. \qquad (3.29)$$

For the special case of wave vector $\mathbf{q} = 0$ this result can be written as

$$\omega_B(0) = \left[\omega_A(2\omega_E + \omega_A)\right]^{1/2}, \qquad (3.30)$$

where we have defined $\omega_A = g\mu_B B_A$ and $\omega_E = 8SJ$ as the bulk anisotropy frequency and exchange frequency, respectively, for this lattice structure. Equation (3.30) then has the form of the well-known AFMR frequency in zero applied field [9]. The surface magnon consists of a single branch at $\omega = \omega_S(\mathbf{q}_\parallel)$ where (see [16])

$$\omega_S\big(\mathbf{q}_\parallel\big) = -(\omega_E/4)\big[1 - \gamma_0^2\big(\mathbf{q}_\parallel\big)\big]$$
$$+ \Big[(\omega_E + \omega_A)\big\{\omega_A + (\omega_E/2)\big[1 - \gamma_0^2\big(\mathbf{q}_\parallel\big)\big]\big\} + (\omega_E/4)^2\big[1 - \gamma_0^2\big(\mathbf{q}_\parallel\big)\big]^2\Big]^{1/2}.$$

$$(3.31)$$

This surface branch occurs at frequencies below the band of bulk magnons, i.e., it is an acoustic surface mode. It is interesting to note that its zero wave-vector limit is

$$\omega_S(0) = [\omega_A(\omega_E + \omega_A)]^{1/2}, \qquad (3.32)$$

which may be compared with equation (3.30). In many bct antiferromagnets the ratio ω_A/ω_E is small (e.g., it is about 0.015 in MnF_2), leading to the conclusion that $\omega_S(0)/\omega_B(0)$ is approximately $\sqrt{0.5}$ ($\simeq 0.71$).

The theory for a semi-infinite sc antiferromagnet, with the structure depicted in figure 3.4(b), is more complicated than the case just described. In part, this is because there is now exchange coupling between sites in the same layer, as well as to adjacent layers, and also each layer contains sites from both sublattices. Therefore there are twice as many equations of motion for the spin amplitudes and the coupling scheme is more intricate. Details of the modified theory are described in [4, 17], so we will quote just the main results for comparison. For the bulk modes it is found that $\omega = \pm\omega_B(\mathbf{q})$, where

$$\omega_B(\mathbf{q}) = \Big[\big(g\mu_B B_A + 6SJ\big)^2 - 4S^2 J^2\big\{\cos(q_x a) + 2\gamma\big(\mathbf{q}_\parallel\big)\big\}^2\Big]^{1/2} \qquad (3.33)$$

and $\gamma(\mathbf{q}_\parallel)$ was defined in equation (3.10). At $\mathbf{q} = 0$ the bulk AFMR frequency takes the same form as equation (3.30) provided we use the appropriate expression $\omega_E = 6SJ$ to define the exchange frequency. If all nearest-neighbour exchange interactions are taken to be the same as the bulk value J, the result for the surface magnon is $\omega = \omega_S(\mathbf{q}_\parallel)$ where

$$\omega_S\big(\mathbf{q}_\parallel\big) = \left[\frac{(\omega_A + \omega_E)\Big\{4\omega_E^2\big[1 - \gamma\big(\mathbf{q}_\parallel\big)^2\big] + 12\omega_A\omega_E + 9\omega_A^2\Big\}}{3(2\omega_A + 3\omega_E)}\right]^{1/2}, \qquad (3.34)$$

which is again an acoustic surface branch. However, if $\mathbf{q}_\parallel = 0$ and if it is assumed that $\omega_A/\omega_E \ll 1$, the above result leads to

$$\omega_S(0) \simeq [\omega_A(2\omega_E + \omega_A/2)]^{1/2}. \qquad (3.35)$$

This frequency for the surface magnon is very close to, but slightly less than, the bulk AFMR frequency in equation (3.30). On the other hand, it is quite different from the result in equation (3.32) for the surface AFMR frequency in the bct case.

We remark that all these results for semi-infinite antiferromagnets become modified if other factors are taken into account, e.g., if the exchange parameters near the surface are varied, if there is a surface anisotropy that is different from the bulk value B_A, or if an applied magnetic field is included.

3.2.3 Exchange-coupled magnetic bilayers

An interesting example where exchange interactions provide a direct coupling across the interface between two exchange-dominated ferromagnets is the Gd/Fe system. This was modelled by Camley [18] as a thin film of Gd on a substrate of bulk Fe. Both materials are ferromagnetic, having an exchange ratio $J_{Fe}/J_{Gd} \sim 6$, but they couple antiferromagnetically across the planar interface with an exchange J_I comparable in magnitude (but opposite in sign) to J_{Fe}. Thus there are competing exchange interactions along with other competing effects if a magnetic field is applied. At low values of magnetic field, applied parallel to the interface, the magnetizations in the two materials are uniform and oppositely directed. At larger field values, however, it was shown that the Fe and Gd spins rotate in the plane of the interface, giving a 'twisted' state for the magnetizations with some broad similarities to the spin-flop phase in simple antiferromagnets (see, e.g., [10]).

A more typical situation involving exchange-coupled bilayers is when the ferromagnetic layers are weakly coupled by exchange across a very thin spacer layer, as depicted in figure 3.5. Usually the spacer is a metal such as Cr with thickness less than a few nanometres. A different exchange mechanism comes into play here, in contrast to the mechanism mentioned in section 3.1 involving localized spins and their overlapping wave functions. Instead, polarization effects in the cloud of conduction electrons in the Cr provide an additional exchange that may extend over greater distances, up to a few nanometres. This is the Rudermann–Kittel–Kasuya–Yosida (RKKY) interaction, which was originally proposed as an indirect exchange mechanism between the $4f$ ion cores in rare-earth metals via the surrounding s electrons. A remarkable property, in addition to the long-range nature, is that it can oscillate with the distance apart of the ions (or with spacer thickness d in the present case) [19]. Long-range coupling across a spacer can also be provided by the magnetic dipole–dipole interactions, as will be considered in the next section.

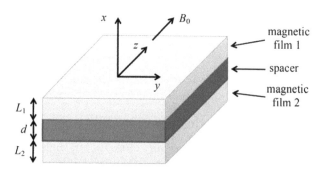

Figure 3.5. Assumed geometry and choice of coordinate axes for two magnetic films separated by a spacer layer.

In work carried out in the late 1980s, that eventually led to the award of a Nobel Prize, Grünberg and Fert separately investigated trilayer systems such as Fe/Cr/Fe in a Cr thickness regime where the Fe layers were coupled antiferromagnetically (see, e.g, [20, 21]). By studying the behaviour of this system in an applied magnetic field using optical and electrical techniques, they discovered the phenomenon of giant magnetoresistance (GMR) which has become of immense technological importance in recording heads and information storage, etc. Parkin *et al* [22], in related experiments also using Fe/Cr/Fe, made a careful characterization of the RKKY coupling across the Cr spacer, confirming that it oscillates with increasing d and eventually decays. Reviews of this topic can be found in, e.g., [23, 24].

As discussed earlier, the usual exchange term in the spin Hamiltonian is proportional to $S_1 \cdot S_2$ for a pair of neighbouring sites labelled as 1 and 2. This type of spin dependence is sometimes called *bilinear exchange*, but other dependences are possible. In particular, there may be exchange contributions proportional to $(S_1 \cdot S_2)^2$ known as *biquadratic exchange*. The total exchange energy acting between two interface spins across the spacer can be written as

$$-J(S_1 \cdot S_2) - J'(S_1 \cdot S_2)^2, \tag{3.36}$$

where J and J' are the bilinear and biquadratic exchange constants. Although it was known to have a small effect in some bulk materials, the biquadratic exchange has been shown to be comparable to the bilinear exchange across metallic spacer layers between some ferromagnets. This was clearly demonstrated for the Fe/Cr/Fe system in experiments by Rührig *et al* [25]. Since J in this system oscillates between positive and negative values as the thickness d is varied, there are some circumstances where $J \approx 0$ and the second term in equation (3.36) dominates. Then if $J' < 0$, which is usually the case for biquadratic exchange, the minimum energy will occur when S_1 and S_2 are perpendicular to one another. This behaviour was confirmed for Fe/Cr/Fe systems in the studies where d was varied. Various explanations have been put forward for the enhanced role of biquadratic exchange at interfaces, including interface roughness and the presence of interface impurities (see, e.g., [26]).

Further discussion of RKKY (bilinear) and biquadratic exchange will come later in this chapter in the context of including the additional role of the dipole–dipole interactions and for multilayers. Finally, we remark that other kinds of interactions across magnetic interfaces have also been investigated. In particular, the 'pinning' of the magnetization in a ferromagnetic film by putting it in direct contact with an antiferromagnet (leading to the phenomenon of *exchange biasing*) can be used as a so-called *spin valve* (see, e.g., [24]).

3.3 Dipolar and dipole-exchange waves in films

We now consider magnetic excitations in the wave-vector regimes that are associated with the dipole-dominated (magnetostatic) and mixed dipole-exchange waves (see figure 3.1). The emphasis will be on ferromagnetic materials, but brief consideration will also be given to antiferromagnets.

3.3.1 Dipole-dominated waves

The usual configuration for considering localized dipole-dominated waves in a ferromagnetic film is shown in figure 3.6, where the static magnetization M_0 is along the z-direction in the yz-plane parallel to the film surfaces. Since the relevant wavelengths are large compared to any lattice constant, a macroscopic or continuum approach is usually followed. In particular, this means that the dipole–dipole interactions can be treated in terms of the Maxwell's equations for the field variables. The first detailed treatment for a thin film or slab with planar surfaces was due to Damon and Eshbach [27], who predicted a surface wave with unusual propagation characteristics. This wave is now generally known as the Damon–Eshbach (DE) wave (see, e.g., [2, 4–6] for reviews). After describing the ferromagnetic case, we give an account of the analogous magnetostatic theory for antiferromagnets.

A. Ferromagnetic case

As a preliminary step to the mode calculation we need to establish a dynamic magnetic susceptibility relationship between the fluctuating magnetization $\mathbf{m}(\mathbf{r})$ and the fluctuating field $\mathbf{h}(\mathbf{r})$, which can be achieved by going back to the vector equation of motion (3.20) and writing it out in component form. Within the linearization approximation used for its derivation, this equation implies $m_z = 0$ while the x- and y-components of the fluctuating fields are related by

$$\begin{pmatrix} m_x \\ m_y \end{pmatrix} = \begin{pmatrix} \chi_a & i\chi_b \\ -i\chi_b & \chi_a \end{pmatrix} \begin{pmatrix} h_x \\ h_y \end{pmatrix}, \tag{3.37}$$

where

$$\chi_a = \frac{\omega_M \omega_0}{\omega_0^2 - \omega^2}, \qquad \chi_b = \frac{\omega_M \omega}{\omega_0^2 - \omega^2}. \tag{3.38}$$

We have defined angular frequencies related to the applied field and static magnetization:

$$\omega_0 = \gamma B_0, \qquad \omega_M = \gamma \mu_0 M_0. \tag{3.39}$$

We note that both of the susceptibility components in equation (3.38) have poles at the FMR frequency ω_0. The 2×2 matrix appearing in equation (5.25) is often referred to as the *gyromagnetic susceptibility tensor* (or matrix).

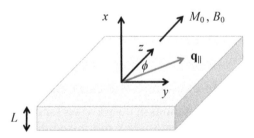

Figure 3.6. Assumed geometry and choice of coordinate axes for dipole-dominated (magnetostatic) and dipole-exchange waves in a ferromagnetic film.

Maxwell's equations, when the electromagnetic retardation effects are negligible, provide the fundamental relationships between $\mathbf{m}(\mathbf{r})$ and $\mathbf{h}(\mathbf{r})$ as [28, 29]

$$\nabla \times \mathbf{m}(\mathbf{r}) = 0, \qquad \nabla \cdot [\mathbf{h}(\mathbf{r}) + \mathbf{m}(\mathbf{r})] = 0. \qquad (3.40)$$

These are the magnetostatic equations and the first one means that a magnetostatic scalar potential $\psi(\mathbf{r})$ can be introduced with the property

$$\mathbf{h}(\mathbf{r}) = \nabla\psi(\mathbf{r}). \qquad (3.41)$$

Then, using equation (5.25) and the divergence condition in equation (3.40), we have for all points within the magnetic material

$$\left(1 + \chi_a\right)\left(\frac{\partial^2\psi}{\partial x^2} + \frac{\partial^2\psi}{\partial y^2}\right) + \frac{\partial^2\psi}{\partial z^2} = 0, \qquad (3.42)$$

whereas outside the magnetic material (where $\chi_a = 0$) we have simply $\nabla^2\psi = 0$.

Assuming now that the magnetic film occupies the region $0 < x < -L$ and a non-magnetic material is outside, we seek solutions for $\psi(\mathbf{r})$ that have the form $\psi_1(x)\exp(i\mathbf{q}_\| \cdot \mathbf{r}_\|)$ in accordance with Bloch's theorem. Here we have denoted $\mathbf{r}_\| = (y, z)$ and $\mathbf{q}_\| = (q_y, q_z) = q_\|(\sin\phi, \cos\phi)$ for the propagation angle ϕ as indicated in figure 3.6. The general solutions for the x-dependence in the three spatial regions (choosing the terms that vanish at $x = \pm\infty$ as required) are

$$\psi_1(x) = \begin{cases} a_1 \exp\left(-q_\| x\right) & \text{for } x > 0, \\ a_2 \exp\left(iq_x x\right) + a_3 \exp\left(-iq_x x\right) & \text{for } 0 > x > -L, \\ a_4 \exp\left(q_\| x\right) & \text{for } x < -L. \end{cases} \qquad (3.43)$$

By using equation (3.42) inside the film it is seen that q_x, which may be real or imaginary, must satisfy

$$(1 + \chi_a)\left(q_x^2 + q_\|^2\right) - \chi_a q_z^2 = 0. \qquad (3.44)$$

The amplitudes in equation (3.43) can be found by applying the usual electromagnetic boundary conditions at the film surfaces $x = 0$ and $x = -L$. This step will determine the physical nature of the modes as bulk-like or localized at the surface(s). In the magnetostatic regime the boundary conditions are first that ψ must be continuous across the film surfaces and second that $(h_x + m_x)$ must be continuous across the film surfaces. The four conditions allow us to solve for the amplitudes and hence to obtain a consistency condition in the form

$$q_\|^2 + 2q_\| q_x(1 + \chi_a)\cot\left(q_x L\right) - q_x^2(1 + \chi_a)^2 - q_y^2\chi_b^2 = 0. \qquad (3.45)$$

Taken together with equation (3.44) for q_x, the above result provides an implicit dispersion relation for the modes.

The results are particularly interesting, as well as being simple, for the case when the propagation angle $\phi = \pi/2$ relative to the magnetization direction, giving $q_z = 0$. This is known as the *Voigt configuration*. When $q_z = 0$ it follows from equation (3.42) that either we have $(1 + \chi_a) = 0$ or $(q_x^2 + q_{\parallel}^2) = 0$. The first condition holds for all real values of q_x provided that $\chi_a = -1$, which has solutions $\omega = \pm\omega_B$ where

$$\omega_B\left(q_x, \mathbf{q}_{\parallel}\right) = [\omega_0(\omega_0 + \omega_M)]^{1/2}. \tag{3.46}$$

This equation gives the frequency of the bulk magnetostatic modes in the Voigt configuration and they are seen to be independent of the wave-vector components. The second condition from above corresponds to q_x being imaginary with $q_x = \pm iq_{\parallel}$. From the form of the solution for the magnetostatic potential given by equation (3.43) in the region $0 > x > -L$, this must correspond to a localized mode near one or other of the surfaces. The solution for the frequency is obtained as $\omega = \omega_S(\mathbf{q}_{\parallel})$ where

$$\omega_S\left(\mathbf{q}_{\parallel}\right) = \left[(\omega_0 + \omega_M/2)^2 - (\omega_M/2)^2 \exp\left(-2q_{\parallel}L\right)\right]^{1/2}. \tag{3.47}$$

This is the DE mode and its frequency increases monotonically as a function of $q_{\parallel} L$. Its limiting values correspond to

$$\omega_S\left(\mathbf{q}_{\parallel}\right) = \begin{cases} [\omega_0(\omega_0 + \omega_M)]^{1/2} & \left(q_{\parallel}L \ll 1\right), \\ (\omega_0 + \omega_M/2) & \left(q_{\parallel}L \gg 1\right). \end{cases} \tag{3.48}$$

Except at $q_{\parallel} = 0$, it follows that we have $\omega_S > \omega_B$ corresponding to an optical surface branch, unlike the situation in the exchange-dominated region. If $q_y > 0$ a calculation of the mode amplitudes shows that the two cases of $q_x = -iq_{\parallel}$ or $+iq_{\parallel}$ correspond to the surface magnetostatic wave being localized near the upper $(x = 0)$ or lower $(x = -L)$ surface, respectively. It is vice versa if $q_y < 0$, but the main point to emphasize is that the surface wave is localized near just one of the two surfaces, instead of near both surfaces as in the exchange-dominated case. This remarkable property is an example of *non-reciprocal propagation*.

The theory may also be worked our for other values of the propagation angle. In particular, when $\phi = 0$ corresponding to propagation parallel or antiparallel to the magnetization direction, it can be shown that there are solutions of equations (3.42) and (3.45) only for q_x real, i.e., there are no surface modes in this case. The equations for the bulk modes can be separated into two types, describing modes that are either symmetric or antisymmetric with respect to the middle of the film:

$$q_x \cot\left(q_x L/2\right) = -q_{\parallel} \quad \text{(symmetric modes),}$$
$$q_x \tan\left(q_x L/2\right) = q_{\parallel} \quad \text{(antisymmetric modes).} \tag{3.49}$$

The discrete real solutions for q_x correspond to a quantization of the bulk modes whose frequencies are now given by

$$\omega_B\left(q_x, \mathbf{q}_{\parallel}\right) = \left[\omega_0\left(\omega_0 + \frac{\omega_M q_x^2}{\left(q_x^2 + q_{\parallel}^2\right)}\right)\right]^{1/2}. \tag{3.50}$$

It is clear from the above expression that the frequency $\omega_B(q_x, \mathbf{q}_\parallel)$ decreases with increasing $q_\parallel$ and so the modes have a negative group velocity $\partial\omega_B/\partial q_\parallel$ parallel to the surface. Hence they are often called *magnetostatic backward bulk modes*. In contrast the group velocity of the bulk modes in the Voigt configuration is zero.

The behaviour for intermediate values of ϕ between 0 and $\pi/2$ is more complicated than for the special cases just described. The main property is that a DE surface mode exists provided ϕ exceeds a critical value ϕ_c given by

$$\phi_c = \left[\frac{\omega_0}{(\omega_0 + \omega_M)} \right]^{1/2}. \tag{3.51}$$

In the physical range the frequency of the DE mode in a thick film (with $q_\parallel L \gg 1$) is

$$\omega_S(\mathbf{q}_\parallel) = [(\omega_0/\sin\phi) + (\omega_0 + \omega_M)\sin\phi]/2, \tag{3.52}$$

which decreases in value as ϕ is reduced from $\pi/2$ to the critical value. The overall behaviour of the bulk and surface modes is shown schematically in figure 3.7.

If similar calculations are made for the magnetostatic modes of a ferromagnetic film with its magnetization perpendicular to the surface [4, 6], it is concluded that there are no surface modes and the bulk-mode frequencies satisfy

$$\omega_B(\mathbf{q}_\parallel, q_z) = \left[\omega_0 \left(\omega_0 + \omega_M - \frac{\omega_M q_z^2}{(q_\parallel^2 + q_z^2)} \right) \right]^{1/2}. \tag{3.53}$$

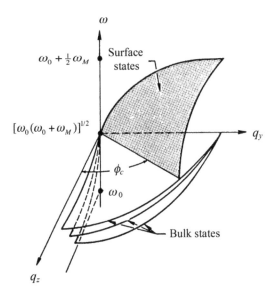

Figure 3.7. Schematic of the dispersion curves for the bulk and surface magnetostatic waves in a ferromagnetic film with static magnetization parallel to the surfaces. The frequency ω is plotted in terms of the in-plane wave-vector components q_y and q_z. After Damon and Eshbach [27].

Note that the z-axis is now perpendicular to the surfaces (since this is the direction of M_0) and so $\mathbf{q}_{\parallel} = (q_x, q_y)$. The quantized values for the perpendicular wave-vector component q_z are obtained from

$$q_z \tan(q_L/2) = q_{\parallel} \quad \text{(symmetric modes)},$$
$$q_z \cot(q_z L/2) = -q_{\parallel} \quad \text{(antisymmetric modes)}. \tag{3.54}$$

The group velocity $\partial\omega_B/\partial q_{\parallel}$ parallel to the surface is positive for these modes and they are often called *magnetostatic forward bulk modes*.

For a double-layer ferromagnetic system we consider the same geometry as in figure 3.5, but with two ferromagnetic films that are now dipolar-coupled to one another across a non-magnetic spacer. For simplicity, we take the two ferromagnets to be of the same material and to have their static magnetizations parallel. Also we assume the special case of the Voigt configuration mentioned earlier. The surfaces of the two films can be taken at $x = 0$ and $x = -L_1$ for the top film and at $x = -(L_1 + d)$ and $x = -(L_1 + L_2 + d)$ for the lower film. The magnetostatic scalar potential satisfies the same equations as before inside and outside the ferromagnets. Thus a set of equations for the x-dependence of the potential can be written down by analogy with (3.43) and the boundary conditions can be applied (see [30] for details). The mode frequencies can eventually be written in the form

$$\omega = [(\omega_0 + \omega_M/2)^2 - (\omega_M/2)^2\alpha)]^{1/2}, \tag{3.55}$$

where $\alpha = 1$ for a bulk mode, giving the same result as in equation (3.46), whereas in the case of a surface mode α satisfies the following quadratic equation:

$$\left[\alpha - \exp(-2q_{\parallel}L_1)\right]\left[\alpha - \exp(-2q_{\parallel}L_2)\right]$$
$$- \alpha\exp(-2q_{\parallel}d)\left[1 - \exp(-2q_{\parallel}L_1)\right]\left[1 - \exp(-2q_{\parallel}L_2)\right] = 0. \tag{3.56}$$

The two solutions for α correspond to the coupled surface magnetostatic modes of the double ferromagnetic film. It is easily verified that in the limit of large separation ($d \to \infty$) the solutions become $\alpha = \exp(-2q_{\parallel}L_1)$ and $\alpha = \exp(-2q_{\parallel}L_2)$, as expected, corresponding to surface modes of the single isolated films. Extensions of the theory have been made to cases where the magnetization directions of the films are not parallel [31] and to a direct interface (i.e., when $d = 0$) between different magnetic materials [32].

B. Antiferromagnetic case

For an extension of the magnetostatic theory to antiferromagnets it is first necessary to obtain the equivalent of equations (5.25) and (3.38) for the gyromagnetic susceptibility. The derivation for a two-sublattice antiferromagnet is given, for example, in [5, 17]. Briefly, the calculation involves writing down equations such as the torque equation (3.17) and the effective magnetization and field, (3.18) and (3.19), for each sublattice individually, including the anisotropy fields. Then the total fluctuating magnetizations from each sublattice are added together and their response to a fluctuating field is deduced. It is found that equation (5.25) still applies, but the expressions for χ_a and χ_b are modified to

$$\chi_a = (\chi^+ + \chi^-)/2, \qquad \chi_b = (\chi^+ - \chi^-)/2, \qquad (3.57)$$

where

$$\chi^{\pm} = \frac{2\omega_A \omega_M}{\omega_A(2\omega_E + \omega_A) - \left(\omega \mp \omega_0\right)^2}. \qquad (3.58)$$

The frequencies ω_A, ω_E and ω were all defined previously and the definition of ω_M is unchanged provided M_0 now refers to the sublattice magnetization. Note that it is only the *static* part of the exchange that is being included through ω_E. Also, in the absence of a static applied field ($\omega_0 = 0$), it follows that $\chi_b = 0$ and the poles of χ_a occur at the AFMR frequency given by $[\omega_A(2\omega_E + \omega_A)]^{1/2}$.

The magnetostatic theory for an antiferromagnetic film can be carried out in a similar manner to the ferromagnetic case, provided equations (3.57) and (3.58) are used and the two-sublattice structure is taken into account [33, 34]. We will quote the results only for the special case of a semi-infinite antiferromagnet in zero applied field. The two branches for the bulk modes are at frequencies $\omega_B^{\pm}(\mathbf{q})$ where

$$\omega_B^+(\mathbf{q}) = [\omega_A(2\omega_E + \omega_A) + 2\omega_A \omega_M \sin^2\theta]^{1/2},$$
$$\omega_B^-(\mathbf{q}) = [\omega_A(2\omega_E + \omega_A)]^{1/2}. \qquad (3.59)$$

Here θ is the angle between the 3D wave vector $\mathbf{q} = (q_x, \mathbf{q}_{\parallel})$ and the in-plane z-axis. At first sight it is surprising that the two bulk magnetostatic frequencies are different, even in zero applied field, unlike the case for the exchange-dominated bulk magnons. This is a consequence of including the lower-symmetry dipolar interactions, which shifts one of the branches (ω_B^+) to a higher frequency than the AFMR frequency. The surface magnetostatic frequency is found to be (see [33])

$$\omega_S(\mathbf{q}_{\parallel}) = \left[\omega_A(2\omega_E + \omega_A) + \frac{2\omega_A \omega_M \sin^2\phi}{1 + \sin^2\phi}\right]^{1/2}, \qquad (3.60)$$

where ϕ is the angle between the 2D wave vector $\mathbf{q}_{\parallel}$ and the z-axis. It can easily be verified that this surface mode satisfies $\omega_B^+(\mathbf{q}) > \omega_S(\mathbf{q}_{\parallel}) > \omega_B^-(\mathbf{q})$, except when angle $\phi = 0$. The surface branch exists as a localized mode for all nonzero ϕ and it satisfies $\omega_S(\mathbf{q}_{\parallel}) = \omega_S(-\mathbf{q}_{\parallel})$. The latter property, which holds only for zero applied field, contrasts with the non-reciprocal propagation behaviour mentioned earlier for surface magnetostatic modes in ferromagnets.

3.3.2 Dipole-exchange waves

The dipole-exchange regime for SWs, as identified in figure 3.1, is more challenging theoretically than the cases treated previously. In part, this is because of the interplay between the long-range nature of the dipole–dipole interactions and the short-range exchange, but perhaps more significantly the boundary conditions at the surfaces and interfaces are more complicated. Two approaches will be presented here, the first being an extension of the macroscopic method just used in subsection 3.3.1. The second is the microscopic (or Hamiltonian-based) method where the dipole–dipole interactions are represented as in equation (3.1) instead of using Maxwell's equations.

A. Macroscopic theory

Here we summarize the formalism as originally developed by Wolfram and Dewames (see [4]) and still widely used. It involves taking into account both the dipolar and exchange terms in the previous equation of motion (3.17), leading to a sixth-order differential equation for the magnetostatic potential. An alternative macroscopic technique was later introduced by Kalinikos and Slavin (see [35]), based on a tensorial Green-function method where the matrix elements of the dipole–dipole interaction are treated perturbatively. The latter is more complicated mathematically, but it has some advantages in yielding approximate analytical results for the linear-response functions needed for applications to inelastic light scattering and SWR measurements from the magnetic modes.

The expression governing the magnetization dynamics can be formed from equations (3.17)–(3.20) by including both the exchange contribution to the total $\mathbf{H}$ field, as done in equation (3.24), as well as the dipolar contribution (denoted by $\mathbf{h}_d$) using the susceptibility relationship in equation (5.25). The result is

$$-i\omega\mathbf{m}(\mathbf{r}) = \hat{\mathbf{z}} \times \left[\omega_M\mathbf{h}_d - \left(\omega_0 - D\nabla^2\right)\mathbf{m}(\mathbf{r})\right], \qquad (3.61)$$

where the SW stiffness constant D is related to the exchange constant as defined earlier. Following now the procedure for the magnetostatic case and adopting the same film geometry as in figure 3.6, we rewrite equation (3.61) in terms of the scalar potential ψ as

$$\left(\hat{w}^2 + \omega_M\hat{w} - \omega^2\right)\left(\frac{\partial^2\psi}{\partial x^2} + \frac{\partial^2\psi}{\partial x^2}\right) + \left(\hat{w}^2 - \omega^2\right)\frac{\partial^2\psi}{\partial z^2} = 0, \qquad (3.62)$$

where we define the differential operator $\hat{w}$ by $\hat{w} = \omega_0 - D\nabla^2$. The above result generalizes equation (3.42) for the purely dipolar case. In the regions outside the film, where $x > 0$ or $x < -L$, we have simply $\nabla^2\psi = 0$.

On putting $\psi(\mathbf{r}) = \psi_1(x)\exp(i\mathbf{q}_\| \cdot \mathbf{r}_\|)$ in accordance with Bloch's theorem, we find that the x-dependence of $\psi_1(x)$ for $x > 0$ and $x < -L$ has the same form as in equation (3.43) whereas inside the film there is a linear superposition of terms:

$$\psi_1(x) = \sum_{j=1}^{3}\left[a_{2j}\exp\left(iq_{xj}x\right) + a_{3j}\exp\left(-iq_{xj}x\right)\right] \quad (0 > x > -L). \qquad (3.63)$$

Here the q_{xj} (with $j = 1, 2, 3$) are the roots of

$$\left(q_x^2 + q_\|^2\right)\left[\left(\omega_0 + Dq_\|^2 + Dq_x^2\right)^2 + \omega_M\left(\omega_0 + Dq_\|^2 + Dq_x^2\right) - \omega^2\right] = 0, \qquad (3.64)$$

where we have simplified by taking the special case of the Voigt geometry ($q_z = 0$ implying $\mathbf{q}_\parallel \perp M_0$). The required roots correspond to $q_{x1}^2 = -q_\parallel^2$ and

$$q_{x2}^2 = \left[\left(\omega^2 + \frac{1}{4}\omega_M^2\right)^{1/2} - \frac{1}{2}\omega_M - \omega_0 - Dq_\parallel^2\right] \bigg/ D,$$

$$q_{x3}^2 = -\left[\left(\omega^2 + \frac{1}{4}\omega_M^2\right)^{1/2} + \frac{1}{2}\omega_M + \omega_0 + Dq_\parallel^2\right] \bigg/ D.$$

(3.65)

It follows that q_{x1} is pure imaginary with the same value as for the surface magnetostatic (or DE) mode and q_{x3} is also pure imaginary, corresponding to another localized mode with no analogue in the dipole-dominated case. On the other hand, q_{x2} is real for most frequencies of interest and it corresponds to a bulk-like mode. Its frequency can be expressed as

$$\omega_B\left(q_{x2}, q_\parallel\right) = \left[\omega_0 + D\left(q_{x2}^2 + q_\parallel^2\right)\right]^{1/2}\left[\omega_0 + \omega_M + D\left(q_{x2}^2 + q_\parallel^2\right)\right]^{1/2}.$$

(3.66)

The statistical weighting (or 'mixing') of these three modes is determined by the boundary conditions, which include the magnetostatic boundary conditions described in subsection 3.3.1. These are insufficient to provide a complete solution for the $\{a_{2j}\}$ and $\{a_{3j}\}$ coefficients appearing in equation (3.63) and so ABCs are required, which can be deduced from the linearized torque equation of motion for the spins at the surface of the ferromagnetic film (see, e.g., [2, 4]). A simplified form commonly adopted for these so-called 'exchange' boundary conditions is

$$\left[\frac{\partial m^\mu(\mathbf{r})}{\partial x} - \xi m^\mu(\mathbf{r})\right]_{x=0} = 0,$$

(3.67)

for the surface at $x = 0$, where ξ is an effective surface anisotropy (or 'pinning') coefficient. There is an analogous condition for the surface at $x = -L$. In the limits of either large or small pinning at each surface it can be shown that there is an approximate quantization of q_{x2} values as $n\pi/L$ with $n = 1, 2, 3, \ldots$, which is consistent with fitting an integer number of half-wavelengths into the film thickness L. The SW spectrum in the dipole-exchange regime then consists roughly of a magnetostatic surface mode with frequency given approximately by equation (3.47) and a series of quantized bulk modes characterized by the different values of the integer n. This description is illustrated by the theory lines in figure 3.8 for the SW frequencies plotted against film thickness L. The bulk-mode frequencies become large for small L because the quantized wave vectors q_{x2} mentioned above become large.

The macroscopic dipole-exchange theory has been extended to antiferromagnetic films [37] by using a two-sublattice model appropriate to bct materials such as MnF_2. It was found that, instead of a cubic equation in q_x^2 being required as in the ferromagnetic case, a fifth-order polynomial is obtained instead (implying a mixing of five waves).

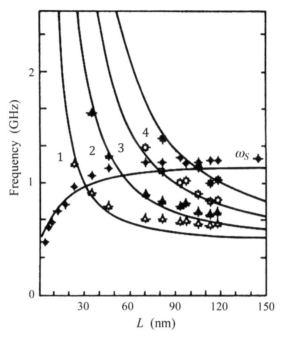

Figure 3.8. Dipole-exchange theory (full lines) for the dependence of the frequencies on film thickness L, showing the surface mode (labelled ω_S) and the lowest standing bulk modes (labelled with their n values). Experimental data points derived from BLS measurements on Fe films are also shown. After Grünberg *et al* [36].

B. Microscopic theory

While a macroscopic approach generally provides a good approximation for the dipole-exchange SWs, a microscopic (or Hamiltonian-based) approach becomes advantageous if the wave vectors become larger and/or if the film thickness is very small corresponding to just a few atomic layers. Following its initial formulation for ultrathin films by Benson and Mills [38], the technique was extended by several authors for ferromagnets (see, e.g., [15, 39]) and antiferromagnets (see, e.g., [40]).

The dipole-exchange spin Hamiltonian can be written as

$$H = -\frac{1}{2}\sum_{i,j}J_{ij}\mathbf{S}_i \cdot \mathbf{S}_j - g\mu_B B_0 \sum_i S_i^z + \frac{1}{2}\sum_{i,j}\sum_{\alpha,\beta}D_{i,j}^{\alpha\beta}S_i^{\alpha}S_j^{\beta}, \qquad (3.68)$$

where we have combined equation (3.1) with (3.3). The final term is the dipole–dipole contribution, where α and β denote x, y or z, and the dipolar coefficient is

$$D_{i,j}^{\alpha,\beta} = \left[|\mathbf{r}_j - \mathbf{r}_i|^2 \delta_{\alpha\beta} - 3\left(r_j^{\alpha} - r_i^{\alpha}\right)\left(r_j^{\beta} - r_i^{\beta}\right) \right] / |\mathbf{r}_j - \mathbf{r}_i|^5. \qquad (3.69)$$

One way to proceed would be to write down the operator equation of motion for S_j^+, as was done in equation (3.7) for the exchange-dominated case. However, we find that the right-hand side involves terms with S^- operators as well as other S^+ operators. Then, in turn, we could write down the equation of motion satisfied by S_j^-, which would involve both S^- and S^+ operators after linearization. This approach is discussed in [17]. An alternative method, which will be useful also when presenting generalizations to *NL* SWs in chapter 7, is to rewrite the spin operators in terms of boson operators. This is motivated by the property that SWs at low temperatures behave like weakly interacting boson particles. The connection can be made in various ways, but the Holstein–Primakoff (HP) transformation [41] is commonly used:

$$S_j^+ = (2S)^{1/2}\left[1 - \left(a_j^\dagger a_j/2S\right)\right]^{1/2} a_j, \qquad S_j^+ = (2S)^{1/2} a_j^\dagger \left[1 - \left(a_j^\dagger a_j/2S\right)\right]^{1/2},$$
$$S_j^z = S - a_j^\dagger a_j. \tag{3.70}$$

The operators for creating and annihilating a boson at site *j* are $a_j^\dagger$ and a_j, respectively. They satisfy the standard quantum-mechanical commutation relations [13]

$$\left[a_i, a_j^\dagger\right] = \delta_{ij}, \qquad \left[a_i^\dagger, a_j^\dagger\right] = [a_i, a_j] = 0, \tag{3.71}$$

and it is a simple exercise, using equations (3.70) and (3.72), to check that the spin commutation relations quoted in equation (3.5) are recovered.

The HP transformation is commonly employed in an approximate form for low temperatures $T \ll T_C$ on the basis that $S^z \simeq S$ which implies that $a_j^\dagger a_j/2S \ll 1$. This property allows us to use the binomial expansion for the square roots in equation (3.70) to give

$$S_j^+ = (2S)^{1/2}\left[a_j - \left(a_j^\dagger a_j a_j/4S\right) + \cdots\right] \tag{3.72}$$

and a similar result for S_j^-. Then the dipole-exchange Hamiltonian in equation (3.68) can be expanded in products of the boson operators to yield

$$H = \text{constant} + H^{(2)} + H^{(3)} + H^{(4)} + \cdots. \tag{3.73}$$

Apart from the constant, which is a contribution to the ground state energy of the ferromagnetic film, each term $H^{(m)}$ involves a product of *m* operators. There is no $H^{(1)}$ term since it vanishes by symmetry in most crystal structures (including cubic), so the lowest-order effect is provided by $H^{(2)}$ and this will be our focus in the following paragraphs. The higher-order terms, starting with $H^{(3)}$ and $H^{(4)}$, describe NL SW processes and will be discussed in chapter 7.

If we now make a 2D Fourier transform from $a_j^\dagger$ in the site representation to $a_n^\dagger(\mathbf{q}_\parallel)$ in a mixed representation with layer number *n* and in-plane wave vector $\mathbf{q}_\parallel$ (just as in subsection 3.2.1 for spin operators), we find a result for $H^{(2)}$ of the form

$$H^{(2)} = \sum_{\mathbf{q}_\parallel, n, m}\left[A_{n,m}^{(2)}\left(\mathbf{q}_\parallel\right)a_n^\dagger\left(\mathbf{q}_\parallel\right)a_m\left(\mathbf{q}_\parallel\right) + \left\{B_{n,m}^{(2)}\left(\mathbf{q}_\parallel\right)a_n^\dagger\left(\mathbf{q}_\parallel\right)a_m^\dagger\left(-\mathbf{q}_\parallel\right) + \text{H.c.}\right\}\right]. \tag{3.74}$$

Here $A^{(2)}$ and $B^{(2)}$ depend on the exchange and dipolar interactions, as well as on the applied magnetic field, and H.c. indicates the Hermitian conjugate. For a model of a thin film with *N* atomic layers their explicit form is given in [15, 42]. It is important to note that, in the absence of dipolar coupling, we have $B^{(2)} = 0$ and so only the first term remains in the above equation. The $A^{(2)}$ coefficients simplify in this case to give consistency with equation (3.9).

Next a linear transformation can be made from the $a^\dagger$ and a operators to a new set of boson operators, which we denote by $b^\dagger$ and b, satisfying boson commutation relations such as those in equation (3.72). The object of making such a transformation (as detailed in [15, 42]) is to 'diagonalize' the linearized Hamiltonian $H^{(2)}$ so that it can be re-expressed as

$$H^{(2)} = \sum_{\mathbf{q}_\|,\ell} \omega_\ell\!\left(\mathbf{q}_\|\right) b_\ell^\dagger\!\left(\mathbf{q}_\|\right) b_\ell\!\left(\mathbf{q}_\|\right). \tag{3.75}$$

The interpretation of this result is that ℓ $(=1, 2, ..., N)$ labels the N discrete SW modes which have frequencies $\omega_\ell(\mathbf{q}_\|)$. Denoting $\mathbf{A}^{(2)}(\mathbf{q}_\|)$ and $\mathbf{B}^{(2)}(\mathbf{q}_\|)$ as the $N \times N$ matrices whose elements are the $A^{(2)}$ and $B^{(2)}$ coefficients appearing in equation (3.74), we find that the SW frequencies are given by the positive solutions (for ω) of

$$\det \begin{pmatrix} \mathbf{A}^{(2)}\!\left(\mathbf{q}_\|\right) - \omega\mathbf{I}_N & 2\tilde{\mathbf{B}}^{(2)}\!\left(\mathbf{q}_\|\right) \\ 2\mathbf{B}^{(2)}\!\left(\mathbf{q}_\|\right) & \mathbf{A}^{(2)}\!\left(\mathbf{q}_\|\right) + \omega\mathbf{I}_N \end{pmatrix} = 0. \tag{3.76}$$

Here the tilde denotes a matrix transpose and $\mathbf{I}_N$ is the $N \times N$ unit matrix. The SW frequencies can be obtained numerically from equation (3.76). An example for an fcc ferromagnetic film is shown in figure 3.9, where the dependence of the frequency on applied magnetic field is shown for the lowest microscopic dipole-exchange mode (solid line).

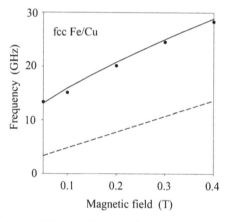

Figure 3.9. Frequency of the lowest SW mode plotted against magnetic field for a 3.8 nm Fe film on a Cu substrate. The in-plane wave vector corresponds to $q_y = 0$ and $q_z = 0.0173\ \mathrm{nm}^{-1}$. The solid line represents dipole-exchange theory for an fcc film with SW propagation parallel to $\mathbf{B}_0$ and the circles are BLS experimental data [43]. For comparison the dashed line shows the theory for the exchange limit. After Nguyen and Cottam [44].

3.4 Experimental results for films and bilayers

There is now an extensive literature covering experimental measurements for SWs in thin films, mainly from inelastic light scattering (BLS and RS) and FMR. Both techniques probe the excitations at long wavelengths, corresponding to the dipole-dominated and dipole-exchange regimes. We mention in particular review articles by Grünberg [45] and Dutcher [46] for detailed accounts.

The first experiments to detect surface magnetostatic modes (the DE modes) were made using direct microwave excitation [47] in a rectangular film of ferrimagnetic yttrium iron garnet (YIG). A pulse from a microwave source at one end of the film induced a surface wave which propagated along one surface of the film until it was registered at a microwave detector placed at the other end of the film. From the measured delay time and the distance between the source and detector, the group velocity v_g of the surface wave can be deduced and compared with the theoretical expression calculated from the dispersion-relation equation (3.47) using $v_g = \partial \omega_S / \partial q_{\parallel}$:

$$v_g = \left(\omega_M^2 L / 4\omega_S \right) \exp\left(-2q_{\parallel} L \right). \tag{3.77}$$

This relationship was tested and verified by making a series of measurements in which the applied field B_0 was varied. Other experiments, in which $q_{\parallel}$ was essentially controlled, provided further evidence for the surface waves.

More convincing evidence for the behaviour of surface and bulk magnetostatic SWs came several years later from BLS measurements on thick platelets of EuO reported by Grünberg and Metawe [48]. By comparing the intensities on the Stokes and anti-Stokes sides of the spectra they gained clear evidence for the non-reciprocal propagation characteristics (mentioned in subsection 3.3.1) of the surface (DE) mode. Subsequently Sandercock and Wettling [49] also used BLS to study the bulk and surface magnetic modes in Fe, where the exchange is much stronger, corresponding to the dipole-exchange regime. Some spectra are shown in figure 3.10 for

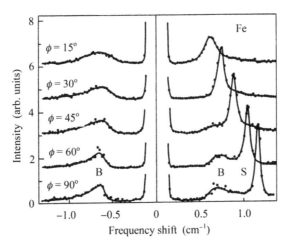

Figure 3.10. Spectra for BLS from the surface of a thick Fe sample with $B_0 = 0.19$T for different propagation angles ϕ. The bulk 'B' and surface 'S' SW peaks are shown. After Sandercock and Wettling [49].

several different values of the propagation angle ϕ (defined in figure 3.6). When $\phi = 90°$ the surface peak 'S' is a strong feature on one side of the spectrum (as a consequence of non-reciprocity), but as the angle is reduced the surface peak shifts to lower frequency, weakens and eventually disappears in accordance with theory. Using equation (3.51) for the critical angle ϕ_c, which is strictly true only in the magnetostatic limit, and the parameters for Fe gives us a reasonable estimate of $\phi_c \simeq 20°$.

Further BLS data, but now for thinner Fe films on different substrates, are included in figures 3.8, 3.9 and 3.11. The bulk modes now become strongly quantized, as discussed earlier, and shift towards higher frequencies at small film thickness L due to exchange terms. These properties are clearly illustrated by the data points in figure 3.8 and are fairly well predicted by the macroscopic dipole-exchange theory. In figure 3.9 where $L = 3.8$ nm it is seen that the BLS data points for the frequency of the lowest bulk mode versus applied field are well described by the microscopic dipole-exchange theory (solid line). The important role of the dipolar interactions, even in this exchange-dominated material, is emphasized by comparison with the dashed line for the exchange limit. Finally, figure 3.11 shows the BLS integrated intensity for the lowest (bulk) SW mode as a function of L for two different choices of substrate material. Although both substrates are non-magnetic they are found to have a large influence on the intensity due to modifications of the SW reflection properties at the Fe–substrate interface.

Turning now to magnetic bilayers, BLS measurements were reported by Grünberg *et al* [36] for Fe films separated by a quartz spacer of variable thickness up to about 100 nm. The behaviour of the coupled modes was found to be in good agreement with the type of theoretical analysis presented in subsection 3.3.1, where the coupling is entirely due to the long-range magnetic dipole–dipole interactions across the spacer. As discussed in section 3.2, however, for certain metallic spacers

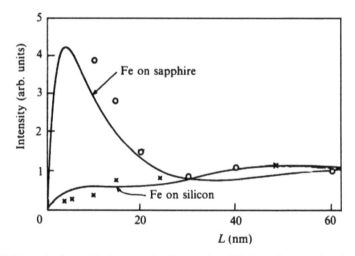

Figure 3.11. BLS intensity for anti-Stokes scattering from surface SWs in Fe films as a function of thickness L for two choices of substrate as indicated. The solid lines are theory and the experimental points (circles and crosses) are due to Grünberg and co-workers. After Cottam [50].

there may be further types of coupling due to RKKY exchange interactions, sometimes occurring together with biquadratic exchange. As mentioned this phenomenon is extensively utilized in devices that depend on GMR. The coupled SW modes can be studied using BLS and/or FMR techniques (see, e.g., [51, 52] for studies of Fe/Cu/Fe and $Ni_{80}Fe_{20}/Ru/Ni_{80}Fe_{20}$ magnetic bilayer films).

3.5 Magnetic wires and stripes

In long wires there is still translational invariance in 1D (taken as the z-direction), so the modes will be characterized by a longitudinal propagation wavenumber q_z and possibly by spatial variations in the transverse xy-plane. Here we consider cases where the cross section of the wire is either circular or rectangular, giving modes of different symmetries.

Magnetic chains (single lines of interacting spins) are a limiting case and they have been a topic of intense interest in the exchange-dominated regime for many years as model theoretical systems for studying quantum phase transitions. On the one hand there may be effects due to competing exchange interactions from the nearest and next-nearest interactions if these have opposite signs. On the other hand there may be anisotropic exchange interactions, such as when the isotropic $\mathbf{S}_1 \cdot \mathbf{S}_2$ interaction between two spins is replaced by a lower-symmetry interaction of the form $\alpha^x S_1^x S_2^x + \alpha^y S_1^y S_2^y + \alpha^z S_1^z S_2^z$, where some or all of the α coefficients differ from unity. Detailed review accounts have been given, for example, by Mikeska and Kolezhuk [53] and Parkinson and Farnell [54]. Our focus, however, will be on quasi-1D wires where there is spatial quantization of the modes in the transverse plane.

3.5.1 Cylindrical wires and tubes

We begin by considering the magnetostatic (dipole-dominated) theory for a long cylindrical wire of a ferromagnetic material with circular cross section of radius R. The calculations are given in detail in [55], so here we will summarize the main steps. Taking the applied magnetic field B_0 to be along the z-axis, which is also the longitudinal axis of the cylinder, we introduce a scalar potential ψ as in subsection 3.3.1, which must again satisfy equation (3.42) inside the material and $\nabla^2 \psi = 0$ outside. We transform from Cartesian (x, y, z) to cylindrical polar coordinates (r, θ, z) and seek separable solutions for ψ of the form $f(r)\exp(im\theta)\exp(iq_z z)$. Here m must be an integer for ψ to be single-valued and the z-dependence is in accordance with the 1D Bloch's theorem. It follows that the radial part $f(r)$ inside the ferromagnet (where $r < R$) must satisfy

$$\left(1 + \chi_a\right)\left(\frac{d^2 f}{dr^2} + \frac{1}{r}\frac{df}{dr} - \frac{m^2}{r^2}f\right) - q_z^2 f = 0, \tag{3.78}$$

where χ_a is the susceptibility term given by equation (3.38). The solutions that are well behaved at $r = 0$ are $f(r) = aJ_m(kr)$, where a is a constant and J_m is a Bessel function of the first kind. The complex coefficient k satisfies

$$\left[\omega_0(\omega_0 + \omega_M) - \omega^2\right]\left(k^2 + q_z^2\right) - \omega_0\omega_M q_z^2 = 0. \tag{3.79}$$

Outside the cylinder (where $r > R$) the solution that is well behaved as $r \to \infty$ is $f(r) = bK_m(kr)$ where K_m is a modified Bessel function of the second kind. The boundary conditions at $r = R$, which have the same form as discussed in subsection 3.3.1, can be used to solve for the coefficients a and b. The results for the magnetostatic modes are as follows. If k is real, equation (3.79) can be re-arranged to give the continuum of bulk modes extending from ω_0 to $[\omega_0(\omega_0 + \omega_M)]^{1/2}$, as previously for a film, with k now playing the role of a radial wave vector. The surface magnetostatic modes, which correspond to k taking discrete imaginary values, have complicated dispersion relations. There are multiple branches, characterized by $m = 1, 2, \ldots$, and at long wavelengths such that $|q_z R| \ll 1$ the surface-mode frequencies simplify to give [55]

$$
\omega_S(q_z) = \begin{cases} \omega_0 + \dfrac{1}{2}\omega_M - \dfrac{1}{4}\omega_M(q_z R)^2\left[0.366 - \ln(q_z R)\right] & (m = 1), \\[2mm] \omega_0 + \dfrac{1}{2}\omega_M - \dfrac{1}{4}\omega_M(q_z R)^2[m^2 - 1]^{-1} & (m > 1). \end{cases}
\tag{3.80}
$$

We note that these surface modes lie above the bulk-mode region and they all become degenerate as $|q_z R| \to 0$.

The incorporation of exchange effects into the above calculation for a macroscopic dipole-exchange theory was made by Arias and Mills [56]. Basically the same procedure as in the film case of replacing ω_0 by the differential operator $\omega_0 - D\nabla^2$ also applies here. This leads to a higher-order differential equation (with the mixing of three waves) for the radial part of the scalar potential ψ. The magnetostatic boundary conditions are supplemented by exchange-dependent ABCs as for the film geometry. A modified form of the above theory was employed by Wang et al [57] to compare with their BLS measurements on cylindrical Ni nanowires with R ranging from 12.5 to 27.5 nm. They deduced that the BLS spectrum should be dominated by standing bulk modes with quantization of the radial wave vector. The approximate values of the mode frequencies, if it is assumed that there is small 'pinning' of the modes at the wire surface, correspond to

$$
\omega = \left\{ D\left(\frac{a_m}{R}\right)^2\left[D\left(\frac{a_m}{R}\right)^2 + \omega_M\right] \right\}^{1/2}, \quad (m = 1, 2, 3, \ldots).
\tag{3.81}
$$

The values of the numerical constants a_m are related to properties of Bessel functions and the three lowest values are $a_1 \simeq 1.84$, $a_2 \simeq 3.05$ and $a \simeq 4.20$. Equation (3.81) holds when $B_0 = 0$ provided $q_z^2 \ll (a_m/R)^2$, which was well satisfied in the experiments. The above simple equation provides excellent agreement with the BLS experimental data, as illustrated in figure 3.12.

Generalizations of the cylindrical wire results to tube geometries, where there is an inner and outer radius to the magnetic material, have also been made for the magnetostatic and dipole-exchange regimes. We mention, in particular, some BLS experimental and theoretical work reported in [58] on Ni nanorings with outer radii of about 50 nm and a wall thickness of about 15 nm. The quantized SWs were detected

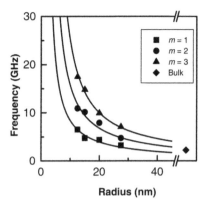

Figure 3.12. Bulk SW frequencies plotted versus radius for Ni cylindrical nanowires in zero applied field. The solid lines are obtained from macroscopic dipole-exchange theory using equation (3.81) for three values of m and the experimental points are deduced from BLS measurements. After Wang *et al* [57].

and studied as a function of a longitudinal applied magnetic field. The accompanying theory was based on using the OOMMF micromagnetic simulation package (briefly mentioned in chapter 1) and the macroscopic dipole-exchange theory.

The microscopic dipole-exchange theory has also been applied to describe the spin dynamics of cylindrical nanowires and is particularly useful when the saturation magnetization is spatially inhomogeneous (e.g., due to edge/interface effects or competing interactions). As an example, we mention theoretical and BLS experimental studies on permalloy ($Ni_{80}Fe_{20}$) cylindrical nanowires [59]. It was found that the presence of a weak single-ion anisotropy in these samples modified the SW properties in a way that contrasted with the behaviour for the Ni studies cited above. In these applications it is convenient and customary to divide the sample up into very small cells (e.g., cubes with sides of length a), which might contain many atomic spins. Then, if a is sufficiently small compared with the length scale associated with fluctuations in the dynamical magnetization, all the atomic spins in the cell will behave as one effective spin and a is an effective lattice parameter. The usual criterion employed is that a should be chosen to be smaller than the material parameter known as the *exchange length*, or more specifically the exchange correlation length, l_{ex}, given by

$$l_{ex} = \sqrt{D/\gamma\mu_0 M_0} \tag{3.82}$$

in terms of the static magnetization M_0 and the SW exchange stiffness parameter D. Physically l_{ex} is the SW wavelength at which the dipolar and exchange contributions in the bulk dispersion relation are comparable in magnitude (see, e.g., equation (3.66)). In permalloy we have $l_{ex} \simeq 5$ nm.

3.5.2 Rectangular wires

Rectangular wire samples are typically fabricated by starting with a thin metallic film of thickness L on a non-magnetic substrate. Then, by lithography and etching, long wires of width W can be formed in arrays with a chosen separation, so that the wires are either independent of one another or are magnetically coupled. Consideration of the latter case will be deferred until section 3.7 on MCs, while we focus here on individual wires. Usually L is a few tens of nanometres while W may be of order 1 μm, so $W \gg L$. Hence these samples may also be referred to as 'stripes' or 'ribbons'.

The pioneering work in studying these types of ferromagnetic nanowires (often formed from permalloy on Si) was carried out by Jorzick *et al* [60], among others, using mainly BLS. This topic has been comprehensively reviewed by Demokritov and co-workers [61, 62]. Some typical experimental data for the SW frequencies as measured by BLS at different values of the in-plane wave vector are illustrated in figure 3.13. The light scattering geometry is such that $\mathbf{q}_{\parallel}$ is in the plane of the original film and perpendicular to the longitudinal axis of the stripe, whereas the magnetization is along the axis of the stripe. Hence this is the analogue of the Voigt configuration for a film, but there is no translational symmetry in the direction of $\mathbf{q}_{\parallel}$. As a consequence, it was argued in [60] that the in-plane wave vector of the DE surface mode, appearing in equation (3.47) for a film, would effectively become quantized as

$$q_{\parallel}^{(n)} = n\pi/W, \quad (n = 0, 1, 2, \ldots), \tag{3.83}$$

which corresponds to fitting an integer number n of half-wavelengths into the stripe width W. The effect of this quantization for the SW frequency is clearly evident at small n in figure 3.13. At larger n the effect is less obvious since the discrete frequencies become closer together and the experimental points lie to a good approximation along the dispersion curve for the DE mode of a complete film (dashed line). The spread of $q_{\parallel}$ values for the SWs with low n was also estimated through mode intensity calculations. Subsequent work focused on improving the

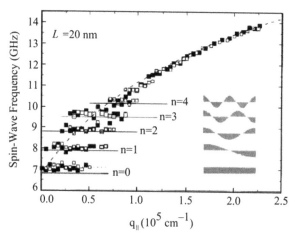

Figure 3.13. SW dispersion for a permalloy nanowire (stripe) of thickness $L = 20$ nm and width $W = 1.8$ μm. The open and closed squares are experimental data points from BLS in the Voigt geometry. The solid horizontal lines are theory and for comparison the DE curve for a complete film with the same L is shown as a dashed line. The mode profiles are depicted as insets alongside the given values of n. After Jorzick *et al* [60].

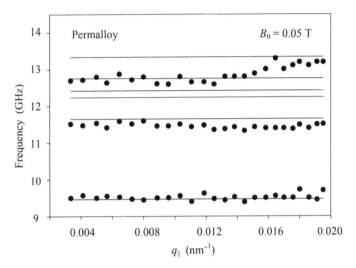

Figure 3.14. Comparison of theoretical (solid lines) and experimental (solid dots) SW dispersion relations for a permalloy nanowire stripe with $L = 20$ nm and $W = 175$ nm. The same geometry is assumed as in figure 3.13. The BLS data are from [66]. After Nguyen *et al* [64].

exchange-type boundary conditions, analogous to equation (3.67), but applied now at the lateral edges of the stripe, e.g., through the concept of dynamical pinning to replace the previous ξ parameter [63].

The microscopic approach, within the same type of dipole-exchange formalism as described in subsection 3.3.2 for films, has also been applied to rectangular nanowires (stripes) [64] with width W much smaller than those in the above paragraph. An example of some calculations, again comparing theory with BLS data in the Voigt geometry, are shown in figure 3.14. The lateral quantization effects are now much more pronounced and the spread of wave vectors is greater. This formalism can be usefully generalized to situations where the static magnetization is spatially inhomogeneous, e.g., to apply to BLS studies of bilayer and trilayer magnetic stripes with dipolar coupling through non-magnetic spacers [65].

3.6 Magnetic superlattices

We now consider SWs in multilayer magnetic systems, in particular superlattices with a periodic or quasi-periodic growth rule. As in earlier sections of this chapter the behaviour is strongly dependent on whether we are dealing with the exchange-dominated regime or whether dipolar effects are important. There are several excellent review accounts giving theoretical and experimental details of this topic; see, for example, [17, 45, 67–69] for mainly periodic superlattices and [70] for quasi-periodic superlattices.

3.6.1 Exchange-dominated waves

We start with the case where the short-range exchange interactions dominate and the alternating media A and B are both sc ferromagnets (assumed for simplicity to have

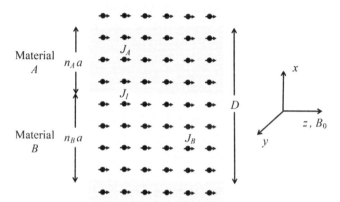

Figure 3.15. Schematic model for a superlattice of alternating layers of two ferromagnetic materials A and B with nearest-neighbour exchange interactions J_A and J_B, respectively, and interface exchange J_I. The layers are stacked in the x-direction and the periodic length D is given by $D = (n_A + n_B)a$ where a is the lattice constant.

the same lattice constant a). The schematic model is depicted in figure 3.15 where the magnetization directions are parallel to the (010) interfaces. The nearest-neighbour exchange interactions and spin quantum numbers are J_A and S_A, respectively, in material A, and likewise J_B and S_B in material B. Nearest neighbours directly across an interface have exchange J_I.

The calculations for an infinitely extended superlattice now proceed using transfer-matrix methods by direct analogy with the cases in section 1.5 (for photonics bands in a bulk-slab model) and section 2.4 (for acoustic phonons or elastic waves in superlattices). The equations of motion for spin operators S_j^+ at each atomic site j are written down and then simplified using the RPA just as in subsection 3.2.1. Expressions for each spin operator are written in terms of forward- and backward-travelling waves (in the x-direction here) by analogy with equations (1.47) and (1.49) and a 2×2 matrix representation can be introduced. The expressions become quite complicated, even for relatively modest values of n_A and n_B. Details can be found in, e.g., [68, 70]. The general behaviour can be anticipated by analogy with the folding of acoustic-phonon modes in chapter 2 (see figure 2.11 and discussion in the text). Here the bulk SWs associated with the constituent films are folded in the reduced Brillouin zone with its edge at $Q = \pi/D$. For waves propagating perpendicular to the interface, the individual film modes extend from frequency $g\mu_B B_0$ to $g\mu_B B_0 + 4S_A J_A$ in material A, and a similar range in material B. This formalism has been modified by Barnás [68] in two respects: first, he extended the derivations to include waves with in-plane wave vector $\mathbf{q}_\parallel = (q_y, q_z) \neq 0$ and, second, he considered semi-infinite superlattices with an outer surface at the terminated end of the stack. In the latter case there is the possibility of surface SW modes of the superlattice when the appropriate modification to Bloch's theorem is made for localized modes. A calculation for the dispersion relations of the bands of superlattice bulk SWs (shaded regions) and discrete surface SWs (dashed lines) is given in figure 3.16 for a semi-infinite superlattice.

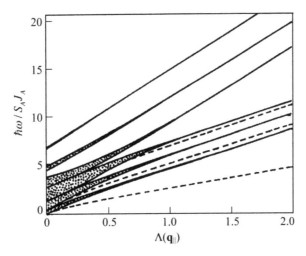

Figure 3.16. Bulk (shaded regions) and surface (dashed lines) exchange-dominated SWs for a semi-infinite super-lattice of alternating ferromagnetic materials A and B with nearest-neighbour exchange interactions. The mode frequency (as the dimensionless quantity $\hbar\omega/S_A J_A$) is plotted in terms of $\Lambda(\mathbf{q}_\parallel) \equiv 1 - \frac{1}{2}[\cos(q_y a) + \cos(q_z a)]$. The assumed parameters are $n_A = n_B = 3$, $J_B/J_A = 2$, $J_I/J_A = 1.4$, $S_B = S_A$ and $B_0 = 0$. There is a B layer assumed at the surface with $J_{BS} = \frac{1}{2}J_A$ for the surface exchange. After Barnás [68].

Various extensions of the work for the exchange-dominated superlattices have been made. For example, some systems will have more complicated ground states for the spin ordering than considered above. Hinchey and Mills [71] considered superlattices with alternating ferromagnetic and antiferromagnetic films, while Camley and Tilley [72] analysed Fe/Gd superlattices where the spins in successive layers have their equilibrium orientations rotated in the plane parallel to the interfaces, leading to a 'twisted phase'. On the experimental side, inelastic neutron scattering was used by Schreyer et al [73] to probe the SWs in short-period Dy/Y magnetic superlattices. These and other developments are described in more detail in the references given earlier in this section.

3.6.2 Dipolar and dipole-exchange waves

The simplest situation in this regime is to consider superlattices where ferromagnetic layers alternate with non-magnetic spacer layers and there is magnetic coupling provided by the dipolar fields across the spacers. The theory for the superlattice modes may be worked out separately for the cases where the magnetization is either parallel [74, 75] or perpendicular [76] to the interfaces.

Here we outline the magnetostatic calculation for a semi-infinite superlattice with in-plane magnetization in the Voigt configuration (see figure 3.17). Then, with the scalar potential written as $\psi_l(x)\exp(i\mathbf{q}_\parallel \cdot \mathbf{r}_\parallel)$ as in subsection 3.3.1, we have in the ferromagnetic layer of the lth cell, i.e., for $-(l-1)D > x > -(l-1)D - d_B$:

$$\psi_l(x) = a_l \exp\{iq_x[x + (l-1)D]\} + b_l \exp\{-iq_x[x + (l-1)D]\}. \quad (3.84)$$

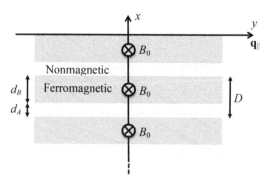

Figure 3.17. Schematic model used for a semi-infinite superlattice in the magnetostatic and dipole-exchange regimes where there are alternating ferromagnetic and non-magnetic spacer layers. The Voigt geometry is assumed in-plane wave vector perpendicular to the applied field direction.

This has a form analogous to that in subsection 1.5.1 for photonic periodic multilayers, where a_l and b_l are constants. The theory proceeds using a transfer-matrix formalism in a very similar manner, in which a_l and b_l in the lth cell define a two-component column matrix. Here q_x has solutions that are discussed in equation (3.44) and the following text. An analogous expression to equation (3.84) can be written down for the potential $\psi_1(x)$ in the non-magnetic spacer region of the lth cell (replacing iq_x by $q_\parallel$ and introducing two new constants). We also require the magnetostatic boundary conditions at each interface and these are the same as discussed in subsection 3.3.1.

By explicit evaluation of the transfer matrix and the use of equation (1.62), we eventually find that the bulk superlattice modes correspond to the solutions, for real values of the Bloch wave vector Q, of

$$\cos(QD) = \cosh\!\left(q_\parallel d_A\right)\cosh\!\left(q_\parallel d_B\right) + g\,\sinh\!\left(q_\parallel d_A\right)\sinh\!\left(q_\parallel d_B\right), \qquad (3.85)$$

where

$$g = 1 + \chi_a - \frac{\chi_b^2}{2(1 + \chi_a)}. \qquad (3.86)$$

Here equation (3.85) is the analogue of the previous Rytov equations (1.64) and (2.39) for optical modes and phonons, respectively. The fact that (3.85) involves the hyperbolic sinh and cosh functions, instead of sin and cos functions as in the previous examples, is a consequence of the superlattice bulk mode being related to a linear combination of the DE *surface* mode for individual films, instead of a combination of bulk modes [68]. The susceptibility terms χ_a and χ_b are defined in equation (3.38).

The possible existence of a surface mode for the semi-infinite superlattice can also be deduced from the above formalism leading to the remarkable conclusion that a surface mode exists only when $d_B > d_A$ (i.e., the magnetic films are thicker than the spacers), in which case its frequency is given by $\omega_S = \omega_0 + \frac{1}{2}\omega_M$. This frequency is the same as the frequency for the DE mode in a thick film and is independent of the values of d_B and d_A. On the other hand, if $d_B < d_A$ there is no superlattice surface mode, because the localization condition for a spatially decaying mode cannot be satisfied. This predicted behaviour was verified in BLS experiments by Grimsditch *et al* [77] and some spectra for Ni/Mo superlattices are shown in figure 3.18. There is an extra peak (the surface mode) observed when $d_B > d_A$; it occurs on only one side of the spectrum, which demonstrates the property of non-reciprocal propagation also predicted by the theory.

When the magnetostatic theory is applied to a ferromagnetic/non-magnetic superlattice in the case where the magnetization M_0 is perpendicular to the interfaces [76], it leads to multiple bands of bulk modes and also, depending on the parameters, multiple surface modes. Non-reciprocal propagation does not occur in this case, which in part is a consequence of the fact that the superlattice modes are now formed from combinations of the *bulk* modes of the magnetic film.

In the dipole-exchange regime for ferromagnetic/non-magnetic superlattices, most of the work has focused on the macroscopic approaches. The first calculations were made by Vayhinger and Kronmuller [78], essentially generalizing the perturbative tensorial Green-function method used for films by Kalinikos and Slavin [35]. Subsequently a more detailed theory was developed by Hillebrands and others (see, e.g., [79]) based on the torque equation-of-motion method for the magnetization described in subsection 3.3.2. One of the challenges in the approach is the proper inclusion of the additional (exchange- and anisotropy-dependent) boundary conditions at the interfaces. Details are given in the review articles on magnetic superlattices mentioned earlier. An example of a dipole-exchange calculation for a finite Co/Pd superlattice with nine periods is given in figure 3.19.

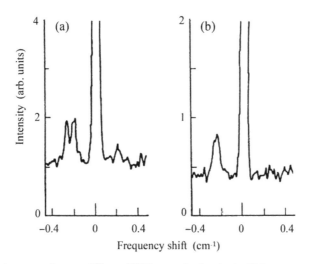

Figure 3.18. Brillouin spectra for two different Ni/Mo superlattices in the Voigt geometry and applied field $B_0 = 0.093$T: (a) $d_A = 8.3$ nm and $d_B = 24.9$ nm; (b) $d_A = 30$ nm and $d_B = 10$ nm. The additional peak due to the superlattice surface mode is in the anti-Stokes (negative shift) side of (a). After Grimsditch *et al* [77].

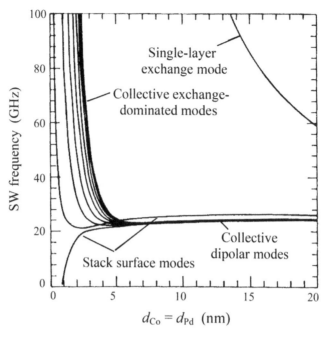

Figure 3.19. Frequencies of dipole-exchange SWs in a Co/Pd superlattice with nine periods calculated as a function of the thickness of the magnetic Co layers, taking $d_{Co} = d_{Pd}$. Other fitting parameters were deduced from BLS measurements. After Hillebrands *et al* [79].

Finally we mention that the experimental and theoretical studies of SWs in quasi-periodic superlattices are described in [70]. Generally the results are more complicated than in the phonon case (see section 2.4) because in the magnetic cases there can be competing interactions that lead to a modified ground state for the equilibrium orientation of the magnetizations, as well as competition between the exchange and dipolar terms.

3.7 Magnonic crystals

In the last decade MCs have become a burgeoning field of study due to their basic physics and their potential for device applications. They consist of lateral arrays of nanowires or nanodots arranged in 1D or 2D with various geometries. The topic was reviewed in a series of articles that appeared in *Journal of Physics D: Applied Physics* in 2010, comprising a general survey [80] and then specialized articles on YIG magnonics [81], BLS studies [82], micromagnetic modelling [83] and logic device applications [84]. More recently there has been a review by Lenk *et al* [85] and a multi-author book edited by Demokritov and Slavin [86]. Most applications are for the dipole-exchange regime of SW behaviour and so that will be our focus here.

3.7.1 Magnonic crystals: theory and experiment

In order to highlight some of the key features of the theoretical methods to incorporate the lateral periodicity of MCs we employ a basic 1D model of an infinitely extended

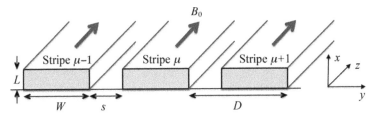

Figure 3.20. Simple geometry for an MC consisting of a 1D array of ferromagnetic stripes of width W and thickness L laterally separated by a non-magnetic spacer of width s. The periodic distance is $D = W + s$.

array of nanowire stripes arranged side-by-side with a non-magnetic intervening material (see figure 3.20). For simplicity the applied field B_0 and hence the static magnetization are taken along the longitudinal z-axis of each stripe. The methods that we outline generalize to other geometric arrangements of nanoelements.

On the macroscopic side, the theories for MCs (as described for example in [82]) have consisted of using micromagnetic simulations, such as adaptations of the OOMMF package mentioned in subsection 1.4.1, or solving analytically a set of dipole-exchange equations analogous to those in subsection 3.3.2 for ferromagnetic films. In the latter case the connection between the fluctuating dipolar field **h** and the fluctuating magnetization **m** can be conveniently expressed in terms of a tensorial Green-function relationship, instead of the previous susceptibility equation (5.25). Following the approach of Kostylev and co-workers [87, 88] the required result can be written in a spatial integral form as

$$\mathbf{h}_{q,Q}(x, y) = \int_A \hat{G}_{q,Q}(x - x', y - y')\mathbf{m}_{q,Q}(x', z')\mathrm{d}x'\mathrm{d}z'. \tag{3.87}$$

Here $\mathbf{h}_{q,Q}$ and $\mathbf{m}_{q,Q}$ are Fourier components of the above fields with respect to the longitudinal wavenumber q in the z-direction and the Bloch wavenumber Q in the y-direction (see figure 3.20). The integration is over the cross-sectional area A of any one stripe in the infinite array and $\hat{G}_{q,Q}$ is a tensorial Green function that has elements given explicitly by

$$\hat{G}_{q,Q}^{\alpha,\beta}(x - x', y - y')$$
$$= \sum_{\mu=-\infty}^{\infty} \exp[\mathrm{i}Q(y - y' - \mu D)]\frac{\partial^2}{\partial\alpha\partial\beta}K_0(q[(x - x')^2 + (y - y')^2]^{1/2}), \tag{3.88}$$

where α and β are Cartesian components. In practice approximate forms of the Green function can be used in order to deduce the magnonic bands representing the behaviour of the SW frequencies as a function of Q [87, 88].

In contrast, in the microscopic dipole-exchange theory for MCs [89], the starting point is a Hamiltonian having exactly the same form as equation (3.68), but now both interwire and intrawire dipole–dipole interactions must be included, as well as the short-range exchange terms. As in the ferromagnetic film case, a transformation can be made to boson operators and then it is the bilinear term $H^{(2)}$ that is of interest in describing the SW modes. By analogy with the representation used for the dipolar fields and Green functions above, a Fourier transform is made for the dipolar coupling term $D_{i,j}^{\alpha,\beta}$, which was defined in equation (3.69), with respect to the longitudinal wavenumber q in the z-direction and the Bloch wavenumber Q in the y-direction. This leads eventually to an expression for $H^{(2)}$ analogous to that found in equation (3.74):

$$H^{(2)} = \sum_{q,Q} \sum_{n,m} \left[A_{n,m}^{(2)}(q,Q) a_n^{\dagger}(q,Q) a_m(q,Q) + \left\{ B_{n,m}^{(2)}(q,Q) a_n^{\dagger}(q,Q) a_m^{\dagger}(-q,-Q) + \text{H.c.} \right\} \right],$$

(3.89)

where $A^{(2)}$ and $B^{(2)}$ can be expressed in terms of the dipole–dipole and exchange parameters, as well as the wave vectors. This equation can then be solved using matrix techniques by following the same formal steps as in the film case [89].

We next present a few examples of experimental results for MCs and their comparison with theory. Some BLS experimental data are shown in figure 3.21, where four magnonic bands were observed with the lower two exhibiting significant dispersion with respect to the Bloch wavenumber Q. The results of macroscopic and microscopic theories are also presented and it is seen that they both lead to satisfactory agreement with experiment. For an example of an MC composed of alternating stripes

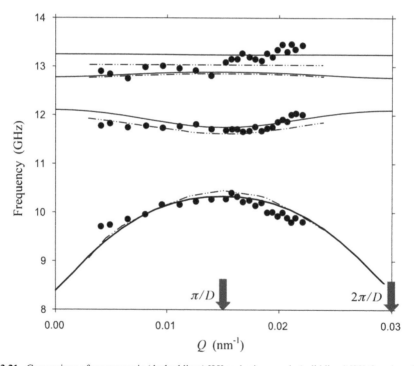

Figure 3.21. Comparison of macroscopic (dashed lines) [82] and microscopic (solid lines) [89] theories with BLS experimental data (black circles) [82], showing SW frequencies versus Bloch wave vector Q for an array of permalloy stripes with $W = 175$ nm, $L = 20$ nm and $s = 35$ nm for the spacer (giving periodic length $D = 210$ nm). After Nguyen and Cottam [89].

of two different ferromagnetic materials (Fe and permalloy) we may refer back to figure 2.17 where there was discussion in the context of the phononic behaviour of this structure. Specifically the lower two panels show BLS spectra of intensity versus SW frequency for two different values of Q. The corresponding magnonic bands deduced from BLS are now displayed in figure 3.22. The blue lines are from micromagnetic simulations and give reasonable agreement with the lowest two bands.

Another type of MC of interest is derived from a periodically modulated waveguide. For example, a single ferromagnetic stripe (with width W large compared to the thickness L) could be considered, where lengthwise modulations of W are introduced. Such a case is illustrated in figure 3.23(a) and this is employed in figure 3.23(b) to make

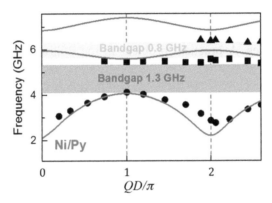

Figure 3.22. Plot of frequency versus reduced wave vector QD/π for the SW bands as deduced from BLS data (see figure 2.17) for a dual phononic/MC composed of alternating Fe and $Ni_{80}Fe_{20}$ (permalloy) nanostripes. After Zhang *et al* [90].

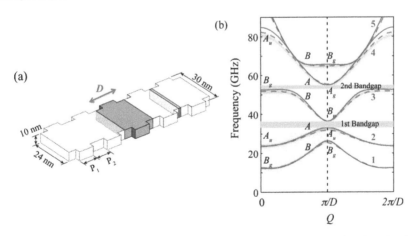

Figure 3.23. MC waveguide consisting of a permalloy stripe with periodic modulations in the width. (a) Geometry and dimensions. The shaded blue block represents the computational unit cell used in the finite-element and microscopic calculations. The green bar indicates where the excitation field is applied in the OOMMF simulations. (b) The magnonic dispersion curves (1–5) plotted versus Bloch wave vector Q. The blue solid, red dashed and green bold lines are data obtained from macroscopic, microscopic and OOMMF calculations, respectively. After Di *et al* [91].

comparisons between the three different theoretical approaches mentioned earlier. This is carried out in a situation where the magnonic bands (labelled A and B in accordance with their symmetry [91]) partially overlap and the band gaps are narrow. Good consistency is demonstrated. Other types of MCs that have been studied include structures that are periodic in the two transverse directions and these are described in the general references given earlier in this section.

3.7.2 Magnon spintronics

By analogy with electronics as being the manipulation of the charge and charge currents of electrons, the term *spintronics* conventionally refers to the control and processing of the electrons through their spins. For example, the manipulation of spin currents in magnetic nanoelements and arrays of these nanoelements, which are composed of metals and semiconductors, offers promising applications for magnetic memories and computational devices. A practical problem in the utilization of spintronics is the relatively short distance (referred to as the spin-diffusion length) over which an electron keeps the memory of its spin orientation. For example, this may be due to the spin–orbit and hyperfine couplings [92]. More specifically, the term *magnon spintronics* has come into usage when concepts from spintronics are used in combination with SW dynamics (see, e.g., [86]) as one way to overcome the limitation mentioned.

A novel device in magnon spintronics has recently been proposed by Chumak *et al* [93]. They employed an MC consisting of a thin film for the ferrimagnet YIG in which an array of parallel grooves had been etched into one surface. The partial reflection and transmission of a SW propagating parallel to the film surface and perpendicular to the grooves constitutes a periodic magnetic system for which the magnonic bands can be calculated using the methods of the previous subsection. The use of this system as a transistor was proposed, with the three terminals represented by a 'source' (a microstrip antenna at one end of the MC providing injection of SWs), a 'drain' (a second antenna acting as a detector of SWs at the other end) and an intermediate 'gate' (a third antenna used to inject another SW signal directly into the device). In this manner, the spin current through the device can be modified by the spin current from the gate. This occurs through an NL process of two interacting SWs scattering into two other SWs (a four-magnon interaction process to be discussed in chapter 7). The spacing of the grooves (representing the periodicity length D) was of order a few hundred micrometres in the reported prototype device [93], but there is the potential to scale down to a few tens of nanometres. The important objective was achieved of having one spin current control another spin current.

References

[1] Keffer F 1966 *Handbuch der Physik* vol 18 ed H P J Wijn (Berlin: Springer) p 1
[2] Gurevich A G and Melkov G A 1996 *Magnetization Oscillations and Waves* (Boca Raton, FL: CRC)
[3] Stancil D D and Prabhakar A 2009 *Spin Waves: Theory and Application* (Heidelberg: Springer)

[4] Wolfram T and Dewames R E 1972 *Prog. Surf. Sci.* **2** 233

[5] Mills D L 1984 *Surface Excitations* ed V M Agranovich and R Loudon (Amsterdam: North-Holland) p 379

[6] Cottam M G (ed) 1994 *Linear and Nonlinear Spin Waves in Magnetic Films and Superlattices* (Singapore: World Scientific)

[7] Hillebrands B and Ounadjela K (ed) 2001 *Spin Dynamics in Confined Magnetic Structures* (Berlin: Springer)

[8] Barman A and Haldar A 2014 *Solid State Phys.* **65** 1

[9] Kittel C 2004 *Introduction to Solid State Physics* 8th edn (New York: Wiley)

[10] Ashcroft N W and Mermin N D 1976 *Solid State Physics* (New York: Saunders College)

[11] Grosso G and Parravicini G P 2013 *Solid State Physics* 2nd edn (Amsterdam: Academic)

[12] Bloch F 1930 *Z. Phys.* **61** 206

[13] Bransden B H and Joachain C J 2000 *Quantum Mechanics* 2nd edn (Englewood Cliffs, NJ: Prentice-Hall)

[14] Cottam M G and Kontos D E 1980 *J. Phys. C: Solid State Phys.* **13** 2945

[15] Costa Filho R N, Cottam M G and Farias G A 2000 *Phys. Rev.* B **62** 6545

[16] Mills D L and Saslow W M 1968 *Phys. Rev.* **171** 488

[17] Cottam M G and Tilley D R 2005 *Introduction to Surface and Superlattice Excitations* 2nd edn (Bristol: IOP)

[18] Camley R E 1987 *Phys. Rev.* B **35** 3608

[19] Hathaway K B 1994 *Ultrathin Magnetic Structures* vol 2 ed B Heinrich and J A C Bland (Berlin: Springer) p 45

[20] Grünberg P, Schreiber R, Pang Y, Brodsky M B and Sowers H 1986 *Phys. Rev. Lett.* **57** 2442

[21] Baibich M N, Broto J M, Fert A, Nguyen van Dau F, Petroff F, Etienne P, Creuzet G, Friederich A and Chazelas J 1988 *Phys. Rev. Lett.* **61** 2472

[22] Parkin S S P, More N and Roche K P 1990 *Phys. Rev. Lett.* **64** 2304

[23] Heinrich B and Bland J A C (ed) 1994 *Ultrathin Magnetic Structures* vol 2 (Berlin: Springer)

[24] Hartmann U (ed) 2000 *Magnetic Multilayers and Giant Magnetoresistance* (Berlin: Springer)

[25] Rührig M, Schäfer R, Hubert A, Mosler R, Wolf J A, Demokritov S and Grünberg P 1991 *Phys. Status Solidi* A **125** 635

[26] Slonczewski J C 1993 *J. Appl. Phys.* **73** 5957

[27] Damon R W and Eshbach J R 1961 *J. Phys. Chem. Solids.* **19** 308

[28] Jackson J D 1998 *Classical Electrodynamics* 3rd edn (New York: Wiley)

[29] Griffiths D J 2012 *Introduction to Electrodynamics* 4th edn (Boston, MA: Addison-Wesley)

[30] Grünberg P 1980 *J. Appl. Phys.* **51** 4338

[31] Grünberg P 1981 *J. Appl. Phys.* **52** 6824

[32] Camley R E and Maradudin A A 1982 *Solid State Commun.* **41** 585

[33] Camley R E 1980 *Phys. Rev. Lett.* **45** 283

[34] Stamps R L and Camley R E 1984 *J. Appl. Phys.* **56** 3497

[35] Kalinikos B A 1994 *Linear and Nonlinear Spin Waves in Magnetic Films and Superlattices* ed M G Cottam (Singapore: World Scientific) p 89

[36] Grünberg P, Cottam M G, Vach W, Mayr C M and Camley R E 1982 *J. Appl. Phys.* **53** 2078

[37] Stamps R L and Camley R E 1987 *Phys. Rev.* B **35** 1919

[38] Benson H and Mills D L 1969 *Phys. Rev.* **178** 839

[39] Stamps R L and Hillebrands B 1991 *Phys. Rev.* B **44** 12417

[40] Pereira J M and Cottam M G 1999 *J. Appl. Phys.* **85** 4949

[41] Holstein T and Primakoff H 1940 *Phys. Rev.* **58** 1098
[42] Nguyen T M and Cottam M G 2005 *Phys. Rev.* B **71** 094406
[43] Tacchi S, Albini L, Gubbiotti G, Madami M and Carlotti G 2002 *Surf. Sci.* **507–510** 324
[44] Nguyen H T and Cottam M G 2011 *J. Phys.: Condens. Matter* **23** 126004
[45] Grünberg P 1989 *Light Scattering* vol 5 ed M Cardona and G Güntherodt (Berlin: Springer) p 303
[46] Dutcher J R 1994 *Linear and Nonlinear Spin Waves in Magnetic Films and Superlattices* ed M G Cottam (Singapore: World Scientific) p 287
[47] Brundle L K and Freedman N J 1968 *Electron. Lett.* **4** 132
[48] Grünberg P and Metawe F 1977 *Phys. Rev. Lett.* **39** 1561
[49] Sandercock J R and Wettling W 1979 *J. Appl. Phys.* **50** 7784
[50] Cottam M G 1983 *J. Phys. C: Solid State Phys.* **16** 1573
[51] Cochran J F, Rudd J, Muir W B, Heinrich B and Celinski Z 1990 *Phys. Rev.* B **42** 508
[52] Liu X M, Nguyen H T, Ding J, Cottam M G and Adeyeye A O 2014 *Phys. Rev.* B **90** 064428
[53] Mikeska H-J and Kolezhuk A K 2004 *Lecture Notes in Physics* vol 645 (Berlin: Springer) chapter 1 pp 1–83
[54] Parkinson J B and Farnell D J J 2010 *Lecture Notes in Physics* vol 816 (Berlin: Springer) pp 1–54
[55] Sharon T M and Maradudin A A 1977 *J. Phys. Chem. Solids* **38** 977
[56] Arias R and Mills D L 2001 *Phys. Rev.* B **63** 134439
[57] Wang Z, Kuok M H, Ng S C, Lockwood D J, Cottam M G, Nielsch K, Wehrspohn R B and Gösele U 2002 *Phys. Rev. Lett.* **89** 027201
[58] Wang Z *et al* 2005 *Phys. Rev. Lett.* **94** 137208
[59] Nguyen T M, Cottam M G, Liu H Y, Wang Z K, Ng S C, Kuok M H, Lockwood D J, Nielsch K and Gösele U 2006 *Phys. Rev.* B **73** 140402(R)
[60] Jorzick J, Demokritov S O, Mathieu C, Hillebrands B, Bartenlian B, Chappert C, Decanini D, Rousseaux F and Slavin A N 1999 *Phys. Rev.* B **60** 15194
[61] Demokritov S O, Hillebrands B and Slavin A N 2001 *Phys. Rep.* **348** 441
[62] Demokritov S O and Hillebrands B 2002 *Spin Dynamics in Confined Magnetic Structures* vol 1 ed B Hillebrands and K Ounadjela (Berlin: Springer) p 65
[63] Guslienko K Y and Slavin A N 2005 *Phys. Rev.* B **72** 014463
[64] Nguyen H T, Nguyen T M and Cottam M G 2007 *Phys. Rev.* B **76** 134413
[65] Gubbiotti G, Nguyen H T, Hiramatsu R, Tacchi S, Madami M, Cottam M G and Ono T 2014 *J. Phys. D: Appl. Phys.* **47** 365001
[66] Gubbiotti G, Nguyen H T, Hiramatsu R, Tacchi S, Madami M, Cottam M G and Ono T 2014 *J. Phys. D: Appl. Phys.* **47** 365001
[67] Mills D L 2000 *Light Scattering in Solids* vol 5 ed M Cardona and G Güntherodt (Berlin: Springer) p 13
[68] Barnás J 1994 *Linear and Nonlinear Spin Waves in Magnetic Films and Superlattices* ed M G Cottam (Singapore: World Scientific) p 157
[69] Hillebrands B 2000 *Light Scattering in Solids* vol 7 ed M Cardona and G Güntherodt (Berlin: Springer) p 174
[70] Albuquerque E L and Cottam M G 2003 *Phys. Rep.* **376** 225
[71] Hinchey L L and Mills D L 1986 *Phys. Rev.* B **33** 3329
[72] Camley R E and Tilley D R 1988 *Phys. Rev.* B **37** 3413

[73] Schreyer A, Schmitte T, Siebrecht R, Bödecker P, Zabel H, Lee S H, Erwin R W, Majkrzak C F, Kwo J and Hong M 2000 *J. Appl. Phys.* **87** 5443
[74] Camley R E, Rahman T S and Mills D L 1983 *Phys. Rev.* B **27** 261
[75] Grünberg P and Mika K 1983 *Phys. Rev.* B **27** 2955
[76] Camley R E and Cottam M G 1987 *Phys. Rev.* B **35** 189
[77] Grimsditch M, Khan M R, Kueny A and Schuller I K 1983 *Phys. Rev. Lett.* **51** 498
[78] Vayhinger K and Kronmüller H 1986 *J. Mag. Mag. Mater.* **62** 159
[79] Hillebrands B, Harzer J V, Güntherodt G, England C D and Falco C M 1990 *Phys. Rev.* B **42** 6839
[80] Kruglyak V V, Demokritov S O and Grundler D 2010 *J. Phys. D: Appl. Phys.* **43** 264001
[81] Serga A A, Chumak A V and Hillebrands B 2010 *J. Phys. D: Appl. Phys.* **43** 264002
[82] Gubbiotti G, Tacchi S, Madami M, Carlotti G, Adeyeye A O and Kostylev M 2010 *J. Phys. D: Appl. Phys.* **43** 264003
[83] Kim S-K 2010 *J. Phys. D: Appl. Phys.* **43** 264004
[84] Khitun A, Bao M and Wang K L 2010 *J. Phys. D: Appl. Phys.* **43** 264005
[85] Lenk B, Ulrichs H, Garbs F and Münzenberg 2011 *Phys. Rep.* **507** 107
[86] Demokritov S O and Slavin A N (ed) 2013 *Magnonics: From Fundamentals to Applications* (Berlin: Springer)
[87] Kostylev M P, Stashkevich A A and Sergeeva N A 2004 *Phys. Rev.* B **69** 064408
[88] Kostylev M P, Schrader P, Stamps R L, Gubbiotti G, Carlotti G, Adeyeye A O, Goolaup S and Singh N 2008 *Appl. Phys. Lett.* **92** 132504
[89] Nguyen H T and Cottam M G 2011 *J. Phys. D: Appl. Phys.* **44** 315001
[90] Zhang V L, Ma F S, Pan H H, Lin C S, Lim H S, Ng S C, Kuok M H, Jain S and Adeyeye A O 2012 *Appl. Phys. Lett.* **100** 163118
[91] Di K, Lim H S, Zhang V L, Ng S C, Kuok M H, Nguyen H T and Cottam M G 2014 *J. Appl. Phys.* **115** 053904
[92] Žutić I, Fabian J and das Sarma S 2004 *Rev. Mod. Phys.* **76** 323
[93] Chumak A V, Serga A A and Hillebrands B 2014 *Nat. Commun.* **5** 4700

IOP Publishing

Dynamical Properties in Nanostructured and Low-Dimensional Materials

Michael G Cottam

Chapter 4

Electronic and plasmonic excitations

In this chapter we examine the properties of a number of electronic-related excitations. We start by considering single-electron states localized at surfaces, followed by the electronic states in 2D graphene sheets and nanoribbons (GNRs). Then we turn our attention to the collective electronic excitations, or plasmons, as part of an analysis of the dispersion properties of the dielectric functions of solids. Some of these topics have already been introduced briefly in chapter 1 and the dielectric properties are utilized further in chapter 5 for the discussion of polaritons.

4.1 Electronic surface states

The possibility of electronic surface states localized at a free planar surface was first investigated in the 1930s in calculations by Tamm [1] and shortly afterwards by Goodwin [2] and Shockley [3]. They all employed a 1D model in terms of the coordinate axis (chosen here as the x-axis) perpendicular to the surface(s). Tamm studied solutions of Schrödinger's equation for an electron in a semi-infinite lattice of square potential wells. In other words, it was the semi-infinite analogue of the Kronig–Penney model described in textbooks on solid-state physics (see, e.g., [4]). In contrast, both Shockley and Goodwin consider finite 1D arrays and their assumed potentials are shown schematically in figure 4.1. The form adopted by Goodwin is more realistic for an atomic potential. A comprehensive review of electric surface states and band-structure theory can be found in [5].

The following outline of the basic theory is based on Shockley's work [3] (but we refer also to the description in [6]). We consider a model with N repeats of a potential well with cells of width d, as in figure 4.1(a). Within any interior cell (with its centre chosen to be at $x = 0$), the general solution of Schrödinger's equation may be written

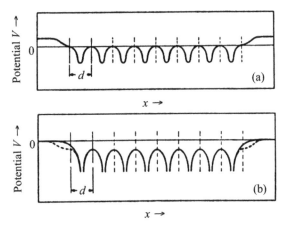

Figure 4.1. Schematic form of the 1D potentials used by (a) Shockley and (b) Goodwin in calculations of electronic surface states. After Shockley [3].

as a linear combination of parts $u(x)$ and $g(x)$ that are, respectively, odd and even functions of x:

$$\psi(x) = a_1 g(x) + i a_2 u(x), \tag{4.1}$$

where a_1 and a_2 are complex coefficients (and the insertion of factor 'i' is for later convenience). If the 1D Bloch's theorem is assumed to hold, then the expression for the wave function in the neighbouring cell centred at $x = d$ is

$$\psi(x) = \exp(iqd)[a_1 g(x - d) + i a_2 u(x - d)]. \tag{4.2}$$

The boundary conditions on the wave function at the $x = d/2$ interface are that ψ and $d\psi/dx$ must be continuous, giving

$$\begin{aligned} a_1 g + i a_2 u &= \exp(iqd)(a_1 g - i a_2 u), \\ a_1 g' + i a_2 u' &= \exp(iqd)(-a_1 g' + i a_2 u'), \end{aligned} \tag{4.3}$$

where u and g, and their derivatives u' and g', are all evaluated at $x = d/2$ and their symmetries have been utilized. The solvability condition for the above two equations in a_1 and a_2 leads to

$$\tan^2(qd/2) = -\left(\frac{g'u}{gu'} \right). \tag{4.4}$$

If we consider first the limit of $N \to \infty$, which is when Bloch's theorem with the use of a real wave vector q becomes valid, equation (4.4) provides us with information about the bulk energy bands in this 1D example. So, for a given $u(x)$ and $g(x)$, if the right-hand side of equation (4.4) is positive then there is real solution for q and the electron energy $E = \hbar^2 q^2 / 2m$ lies within a pass band of electronic states (denoting m = electronic mass). A band gap (or stop band) corresponds to when the right-hand side of equation (4.4) is negative.

By analogy with analogous calculations made in chapters 2 and 3 for other excitations in 1D, the possibility of complex solutions for q arises when N is finite.

This situation can be investigated by matching the Bloch wave functions inside the material to the wave function in the outside regions, where the potential can be assumed to have a constant value V_0. Hence the exterior wave functions will have the form of $\exp(\kappa x)$ on the left as $x \to -\infty$ and $\exp(-\kappa x)$ on the right as $x \to \infty$, where κ is a real and positive decay constant:

$$\kappa = \sqrt{2m(V_0 - E)/\hbar^2}. \qquad (4.5)$$

The outcome of treating the finite-N case in this manner is that the continuum of energy states in the pass bands now consists of discrete ('quantized') levels and also surface branches might occur in the stop bands, depending on details of the modelling. A numerical example, as obtained by Shockley [3], is shown in figure 4.2 for $N = 8$. When the cell size d is very large, the energy-level diagram consists simply of the three levels (each one being eight-fold degenerate) for the isolated atoms. As d is reduced to a value comparable with κ^{-1}, as characterized by d_1 in the figure, the degeneracy is lifted and the eight discrete levels in each band become evident. Then for d much smaller than κ^{-1}, as represented by d_2, it is seen that surface branches may occur in the band-gap regions.

Other similar analytic 1D calculations are presented in the books by Lüth [5], Zangwill [7], and Lannoo and Friedel [8]. They also provide details of generalizations for surface states in the formalism of electron-band theory applied to 3D structures, including schemes of surface reconstruction (where appropriate). Electronic states that are localized with interfaces between different materials are also analysed.

A distinction is often usefully made between *intrinsic* and *extrinsic* electronic surface states. The former are those described so far in this section, i.e., they are a property of microscopically clean surfaces (even if surface reconstruction of the atoms and bonds may have taken place, as in pure Si surfaces [8]). In contrast, extrinsic surface states are associated with adsorbed or absorbed atoms or molecules and have also been widely studied both theoretically and experimentally [8].

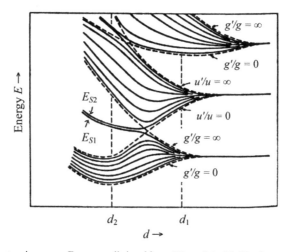

Figure 4.2. Plot of electronic energy E versus cell size d for a 1D model with $N = 8$ atomic potential wells. The dashed lines represent the boundaries of the bands corresponding to $N \to \infty$. After Shockley [3].

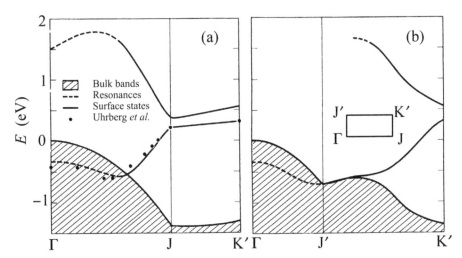

Figure 4.3. Dispersion of surface electronic states for Si(111) with a 2 × 1 reconstruction. (a) Comparison of theory (full line) and ARPES measurements (data points) for Γ–J–K′. (b) Theory for Γ–J′–K′. The labelling of the 2D surface Brillouin zone is shown as an inset in (b). After Northrup and Cohen [9].

An interesting case is provided by the (111) surface of Si, which can undergo a variety of surface reconstructions in both the pure state, when cleaved from a bulk Si crystal, or when the surface is patterned by a chosen coverage of (e.g.) Al, Ga or In. As an example, we show in figure 4.3 a comparison of theory with experiment for a clean Si(111) surface with a 2 × 1 reconstruction [9], meaning that the lengths of the basic vectors (a_1 and a_2 in the notation of subsection 1.1.1) are scaled by factors 2 and 1, respectively, in the reconstructed 2D surface layer. There are two cases shown for the energy E of the surface states, depending on the directions chosen for the wave vectors in the 2D Brillouin zone (see the inset for the standard notation). In figure 4.3(a) some experimental data deduced from angle-resolved photoemission spectroscopy (ARPES) are indicated, giving reasonable agreement with theory. Some more recent references relating to surface electronic states in semiconductor structures including some wires and dots, and mostly in the context of photonics applications, can be found in [10].

4.2 Graphene sheets and ribbons

Graphene is an ML or 2D form of carbon that has been intensively studied, especially in the last decade, for its remarkable electronic band structure including zero-gap features. It shows amazing versatility in its electronic properties depending on whether it is produced in a pristine form or whether edges, impurities, adatoms, vacancies (or other forms of defects) are introduced. Likewise, the tailoring of magnetic properties (e.g., at edges) has attracted an enormous amount of attention. Although graphene was not successfully isolated and fabricated experimentally until 2004 in work [11] that led to the awarding of the 2010 Nobel Prize in Physics to Geim and Novoselov, there were some theoretical investigations long beforehand (see, e.g., [12]). Some excellent reviews of this topic have been provided by Castro-Neto *et al* [13] and Power and Ferreira [14].

The lattice structure of graphene was discussed already in chapter 1 (see figure 1.2) in terms of a 2D hexagonal or honeycomb space lattice with two interpenetrating sublattices of carbon atoms, labelled as A and B. Choices for the basic lattice vectors of a unit cell with two carbon atoms per cell were given, as well as the corresponding basic vectors of the reciprocal lattice. The σ bond between adjacent carbon atoms that are separated by a distance $a_0 = 0.142$ nm results from the sp^2 hybridization. Additionally there is also a weaker π bonding between electrons with delocalized wave function perpendicular to the planar structure and it is these π electron properties that are conventionally studied for the low-lying electronic excitations. The edge states of graphene have been of particular interest following seminal results reported in [15, 16]. Apart from the edges being of intrinsic novelty they can also be functionalized by other chemicals.

GNRs can be envisioned as long structures having a pair of edges along the same chiral direction in the bipartite lattice at a finite width apart. The simplest cases, known as zigzag and armchair GNRs are depicted in figure 4.4(a) and (b) respectively. Their differences can be highlighted by considering the rows of atoms formed by and parallel to the edges. The rows can be labelled by integer n ($= 1, 2, 3, \ldots N$), as in the zigzag case shown here for $N = 8$. In the zigzag case each row contains only A or only B atoms, whereas each row in the armchair case contains equal numbers of A and B atoms. Also the spacing of the rows follows a different scheme in the two cases.

A calculation of the bulk and edge modes in these GNRs will now be outlined, showing that the latter modes have quite distinctive properties. It is the zigzag case

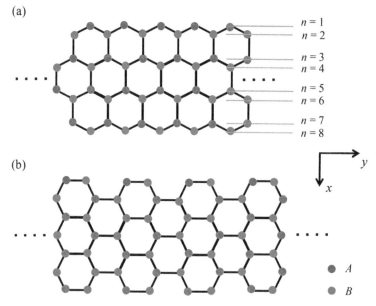

Figure 4.4. Geometry of GNRs with (a) zigzag edges and (b) armchair edges. Each carbon atom lies either on sublattice A or B as indicated. The GNRs are finite in the x-direction and are assumed to be infinite in the y-direction.

that is of greater interest for the edge modes [15, 16] and so we focus on that structure. The starting point is the Hamiltonian, which we take to have the electronic tight-binding form in second quantization as

$$H = -\sum_{i,j} t_{ij}\left(a_i^+ a_j + a_i a_j^+\right),$$ (4.6)

where the fermion operators denoted by a_i^+ and a_i create and annihilate, respectively, an electron at any site i, corresponding to an A sublattice site if i is located on an odd n row and a sublattice B site if i is on an even n row. Due to their different statistics, the fermion operators satisfy anticommutation properties among themselves [17], instead of the commutation properties used for boson operators in the previous chapter. Also t_{ij} is the hopping energy for an electron to move to any of the nearest neighbours, which are on opposite sublattices (in fact, on adjacent rows). This hopping energy in graphene is denoted by t (and is known [13] to be $\simeq 2.8$ eV), except possibly at the zigzag edge where we assume that it may take a modified value denoted by t_e. We may now make a Fourier transform in the usual manner from site labels to a representation with 1D wave vector q_y along the y-direction and row number n. Equation (4.6) can be rewritten as

$$H = -\sum_{q_y,n,n'}\left[\tau_{nn'}(q_y)a_{q_y,n}^+ a_{q_y,n'} + \tau_{nn'}(-q_y)a_{q_y,n}a_{q_y,n'}^+\right],$$ (4.7)

where the transformed hopping factor $\tau_{nn'}(q_y)$ for the zigzag structure has the form

$$\tau_{nn'}(q_y) = t\left[\beta(q_y)\delta_{n',n\pm 1} + \delta_{n',n\mp 1}\right].$$ (4.8)

The upper (lower) signs refer to n odd (even) and we have defined

$$\beta(q_y) = 2\cos\left(\frac{\sqrt{3}}{2}q_y a\right).$$ (4.9)

The excitations can now be investigated by using either the operator equation of motion, as in (1.36), or the Green-function equation of motion, as in (A.22). Both involve working out the commutators of the creation or annihilation operators with the above Hamiltonian [18, 19]. However, the latter method (which is outlined in a general way in section A.2 of the appendix) is preferable here, since it eventually provides fuller information about the localized edge modes. From equations (4.7) and (A.22) we deduce that

$$\omega\left\langle\left\langle a_{q_y,n}^+; a_{q_y,n'}\right\rangle\right\rangle_\omega = \frac{1}{2\pi}\delta_{n,n'} + \sum_m \tau_{nm}(q_y)\left\langle\left\langle a_{q_y,m}^+; a_{q_y,n'}\right\rangle\right\rangle_\omega.$$ (4.10)

This expression can be recognized as equivalent to a set of finite-difference equations in terms of integers n and n'. For any given value of n' the finite-difference equations for variable n can be reorganized such that the odd-n terms (referring to sublattice A) are eliminated in terms of the even-n terms (sublattice B). The resulting equations can then be recast in a matrix form as $\mathbf{A}\,\mathbf{G}_{n'} = \mathbf{b}_{n'}$, where $\mathbf{G}_{n'}$ is the column matrix of

Green functions $\langle\langle a^+_{q_y,n}; a_{q_y,n'}\rangle\rangle_\omega$ with n even, $\mathbf{b}_{n'}$ is another column matrix with elements defined by

$$(\mathbf{b}_{n'})_n = \frac{\omega}{2\pi t^2 \beta(q_y)}\delta_{n,n'}, \tag{4.11}$$

and $\mathbf{A}$ is a TDM that can be separated into two parts as $\mathbf{A}_0 + \mathbf{D}$ as described in section A.1.

To simplify matters, if we take the case of a wide GNR (the semi-infinite limit where $N \to \infty$) then $\mathbf{A}_0$ takes precisely the form of equation (A.3) with

$$d = \left[(\omega/t)^2 - \beta^2(q_y) - 1\right]\big/\beta(q_y), \tag{4.12}$$

and the edge perturbation matrix $\mathbf{D}$ has only one nonzero element given by

$$D_{1,1} = D_e \equiv \left(1 - \frac{t_e^2}{t^2}\right)\beta(q_y). \tag{4.13}$$

We note that the edge parameter D_e vanishes if $t_e = t$ for the hopping terms.

It is now straightforward to solve for the Green functions and hence for the frequencies of the excitations, by following the approach in section A.1 (see also [19]). In particular, it is helpful to introduce the complex variable x, which is formally related to the frequency ω through equations (4.12) and (A.5). Then equation (A.13) is utilized. To simplify the algebra we will quote the results only for the special case where one of the Green-function labels refers to the zigzag edge (i.e., we choose $n' = 1$):

$$\left\langle\left\langle a^+_{q_y,1}; a_{q_y,1}\right\rangle\right\rangle_\omega = \frac{t_e^2\beta(q_y)}{2\pi t^2\omega(1 + xD_e)}x \quad (n = 1),$$

$$\left\langle\left\langle a^+_{q_y,n}; a_{q_y,1}\right\rangle\right\rangle_\omega = \frac{t_e\left[\beta(q_y) + x^{-1}\right]}{2\pi t\omega(1 + xD_e)}x^{(n+1)/2} \quad (\text{odd } n > 1), \tag{4.14}$$

$$\left\langle\left\langle a^+_{q_y,n}; a_{q_y,1}\right\rangle\right\rangle_\omega = \frac{t_e}{2\pi t^2(1 + xD_e)}x^{n/2} \quad (\text{even } n).$$

There are several points to make regarding the physical interpretation of these rather formal expressions. First, each Green function involves overall factors of x to a positive integer power. As explained in section A.1 for the TDM method, the bulk-like (or propagating) modes are associated with $|x| = 1$. In fact, if we now put $x = \exp(i3q_xa/2)$ as a phase factor and use equations (4.12) and (A.5) we deduce that $\omega = \omega_B$ where

$$\omega_B(q_x, q_y) = \pm t\sqrt{\beta^2(q_y) + 2\beta(q_y)\cos(3q_x/2) + 1}. \tag{4.15}$$

This is identical to the standard form of the dispersion relation for the electronic band in an infinite graphene sheet [13], where q_x is a real wave-vector component in the x-direction. By contrast, any poles of the above Green functions with $|x| < 1$ will correspond to localized modes, i.e., edge modes in this graphene example. One way in which this can occur is through the factor $(1 + xD_e)$ appearing in all cases in equation (4.14), so we conclude that there is an edge mode for $x = -1/D_e$, provided the localization condition that $|D_e| > 1$ is satisfied. Solving for the dispersion relation then gives $\omega = \omega_e$ where

$$\omega_e(q_y) = \pm t \sqrt{\left(t_e^2/t^2\right)\beta^2(q_y) + t_e^2 \Big/ \left(t_e^2 - t^2\right)}. \qquad (4.16)$$

The localization condition for this type of edge mode can be satisfied *either* if $0 < (t_e/t)^2 < 0.5$ (when $D_e > 1$) *or* if $(t_e/t)^2 > 1.5$ (when $D_e < -1$). These represent, respectively, acoustic and optic edge modes, as illustrated by the numerical examples in figure 4.5. We note, in particular, that there are no edge modes of this type if $t_e = t$, so they arise specifically as a consequence of perturbations in the edge hopping.

There is, however, another more subtle edge-mode effect that arises only from the odd-n terms in equation (4.14) for the semi-infinite GNR. It can be seen that these Green functions also have poles (vanishing denominators) at $\omega = 0$ and this is the case even when $t_e = t$. A simple rearrangement of equations (4.12) and (A.5) leads to to an expression for ω^2 in terms of x as

$$\omega^2 = t^2 \Big[\beta(q_y) + x\Big]\Big[\beta(q_y) + x^{-1}\Big]. \qquad (4.17)$$

This shows that $\omega = 0$ implies that either $x = -\beta(q_y)$ or $x = -1/\beta(q_y)$. In fact, it is the former condition that corresponds to an edge mode, as can be seen by rewriting the second line of equation (4.14) in the form

$$\left\langle\left\langle a_{q_y,n}^+; a_{q_y,1}\right\rangle\right\rangle_\omega = \frac{t_e}{2\pi t^2(1 + xD_e)}\sqrt{\frac{\beta(q_y) + x^{-1}}{\beta(q_y) + x}}\, x^{(n+1)/2} \quad (\text{odd } n > 1). \quad (4.18)$$

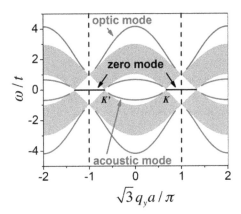

Figure 4.5. Dispersion curves for the mode frequencies versus longitudinal wavenumber q_y for a semi-infinite GNR with a zigzag edge. The two regions of bulk modes are shown shaded, with K and K' indicating the Dirac points where there is zero gap (see the text). The vertical dashed lines correspond to the boundaries of the first Brillouin zone. Examples of optic edge modes when $t_e/t = 1.6$ and acoustic edge modes when $t_e/t = 0.4$ are shown, as well as the zero mode. After Akbari-Sharbaf and Cottam [19].

We therefore conclude that this type of edge mode exists only on the *odd* rows of the semi-infinite GNR. Also the localization condition must be satisfied that $|\beta(q_y)| < 1$, i.e., q_y must lie in the range $2\pi/3\sqrt{3} < |q_y a| < \pi/\sqrt{3}$ in the first Brillouin zone. This excitation is often referred to as the *zero edge mode* and it was first predicted by Fujita *et al* [15]. It is seen as the flat line in figure 4.5 extending from the Dirac points (labelled K and K') to the Brillouin-zone boundary. The above analysis can be extended to show that the conclusions for the frequency and existence of the zero mode still apply when $t_t \neq t$, but the spectral intensities (coming from the Green functions) are modified. The occurrence of these Dirac points, corresponding to 2D wave vectors where the electronic spectrum is gapless, is one of the properties that makes graphene of such special interest (see [13]). The wave vectors for K and K' in our notation are $(q_x, q_y) = (2\pi/3a, \pm 2\pi/3\sqrt{3}\,a)$. For an excitation close to the K point we can easily deduce from equation (4.15) that the approximate bulk-like dispersion relation is

$$\omega_B\big(q_x, q_y\big) \simeq \pm t\sqrt{\delta_x^2 + 3\delta_y^2} \quad \big(|\delta_x| \ll 1, |\delta_y| \ll 1\big), \tag{4.19}$$

where $\delta_x = \pi - \frac{3}{2}q_x a$ and $\delta_y = \frac{\pi}{3} - \frac{\sqrt{3}}{2}q_y a$ are the rescaled dimensionless wave-numbers used in making the expansions.

An analogous analysis can be made for the excitations associated with a semi-infinite GNR having an armchair edge, as illustrated in figure 4.4(b). It is found that the bulk modes can again be described by equation (4.15), *provided* that the roles of wave-vector components q_x and q_y are formally interchanged (so that q_y still denotes the longitudinal wave vector). In this case the zero-gap point in the bulk spectrum occurs at $q_y = 0$. However, the most important difference applies to the edge modes. It is found that there is no analogue of the zero mode for armchair edges. Further discussion is provided in [16, 20], including examples of ribbon edges with different orientations and GNRs with finite width.

The experimental evidence for edge modes in GNRs is indirect, coming from their influence on the electronic density of states which can be probed by electron spin resonance from any paramagnetic impurity centres at the zigzag edges [21]. Lines of impurities or defects, along either the zigzag or armchair direction, are also of interest in GNRs. We mention that topological line defects are discussed, for example in [22], while self-assembly of lines of metal atoms on graphene is explored in [23].

4.3 Bulk dielectric functions

In subsection 1.1.2 we introduced the concept of the (linear) dielectric function, written either as $\varepsilon(\mathbf{q}, \omega)$ or more simply as $\varepsilon(\omega)$ if spatial dispersion is negligible. Also a simple expression for $\varepsilon(\omega)$ was quoted in equation (1.21) for an electron-gas plasma and this was employed to illustrate the concept of a plasmon polariton. We now present an extended discussion of the dielectric function for a range of different types of materials and excitations, including some cases of spatial dispersion. Further accounts of this topic can be found, for example, in [6, 24–26].

We begin with a re-examination of the dielectric behaviour of an electron gas as in a metal or a doped semiconductor. A full quantum-mechanical treatment can be made in terms of the Lindhard function [27], but here we will follow a more straightforward physical approach where the electron gas is described hydrodynamically as a continuous fluid of electrons [25] with particle density $n(\mathbf{r}, t)$, velocity $\mathbf{v}(\mathbf{r}, t)$ and charge current density $\mathbf{j}(\mathbf{r}, t) = -n(\mathbf{r}, t)e\,\mathbf{v}(\mathbf{r}, t)$. The continuity equation of hydrodynamics requires that

$$\partial n/\partial t + \nabla \cdot (n\mathbf{v}) = 0, \tag{4.20}$$

and the velocity obeys the acceleration equation in the form

$$m\frac{D\mathbf{v}}{Dt} = -e(\mathbf{E} + \mathbf{v} \times \mathbf{B}) - \frac{m\mathbf{v}}{\tau} - \left(\frac{m\beta^2}{n}\right)\nabla n. \tag{4.21}$$

Here we have employed the total derivative $D/Dt = \partial/\partial t + \mathbf{v} \cdot \nabla$ and we have included electric- and magnetic-field terms for the force on an electron. Damping is added through the term involving the collision time τ and the final term allows for the compressibility of the electron gas, where it is often assumed that $\beta^2 = 3v_F^2/5$ with v_F denoting the Fermi velocity [25].

The next step is to linearize the above expressions by taking the electric field as the driving term, so $\mathbf{E}$ and $\mathbf{v}$ are first-order quantities and we write the fluctuating density as $n = n_0 + n_1$, where n_0 is a constant. Also we assume that the time-dependent quantities vary as $\exp(-i\omega t)$, giving the linearized equations as $\mathbf{j} = -n_0 e\mathbf{v}$ and

$$-i\omega n_1 + n_0 \nabla \cdot \mathbf{v} = 0, \tag{4.22}$$

$$i\omega m\mathbf{v} = e(\mathbf{E} + \mathbf{v} \times \mathbf{B}_0) + m\mathbf{v}/\tau + (m\beta^2/n_0)\nabla n_1. \tag{4.23}$$

The term with B_0 in equation (4.23) allows for the effect of a static magnetic field in addition to the electric field. Eliminating n_1 yields

$$i\omega m\mathbf{v} = e(\mathbf{E} + \mathbf{v} \times \mathbf{B}_0) + m\mathbf{v}/\tau - (im\beta^2/\omega)\nabla\nabla \cdot \mathbf{v}, \tag{4.24}$$

from which we can now deduce the dielectric function. Taking $\mathbf{E}$ to have a plane-wave form proportional to $\exp(i\mathbf{q} \cdot \mathbf{r})$ in a bulk material, and putting $\mathbf{B}_0 = 0$ for the moment, we find

$$\mathbf{j} = \left(\frac{ie^2 n_0}{m}\right)\frac{\mathbf{E}}{\omega^2 + i\omega/\tau - \beta^2 q^2}. \tag{4.25}$$

From Maxwell's equation for curl $\mathbf{H}$, when expressed in two equivalent forms, we have

$$i\mathbf{q} \times \mathbf{H} = \mathbf{j} - i\omega\varepsilon_0\varepsilon_\infty \mathbf{E} = -i\omega\varepsilon_0\varepsilon(\mathbf{q}, \omega)\mathbf{E}. \tag{4.26}$$

Here the constant ε_∞, which may differ from unity, has been included to allow for the effects of excitations with higher frequencies than the plasma mode. The modification is typically unimportant for metals, but may be useful for doped semiconductors. From equations (4.25) and (4.26) we arrive at

$$\varepsilon(\mathbf{q}, \omega) = \varepsilon_\infty \left(1 - \frac{\omega_p^2}{\omega^2 + i\omega/\tau - \beta^2 q^2} \right) \tag{4.27}$$

in terms of the plasma frequency ω_p defined earlier. This represents the dielectric function for an electron-gas plasma with effects of both spatial dispersion (q-dependence) and damping included. It is a generalization of the simpler result quoted in equation (1.21) and plotted in figure 4.6(a).

The extension to retain the effects of a static applied magnetic field, by using the full form of equation (4.24), is straightforward but algebraically more lengthy. In fact, the dielectric function is found to generalize from a scalar to a gyrotropic matrix (or tensor) quantity. If the applied magnetic field is along the z-direction and spatial dispersion is ignored, the required expression is

$$\bar{\varepsilon} = \begin{pmatrix} \varepsilon_a & i\varepsilon_b & 0 \\ -i\varepsilon_b & \varepsilon_a & 0 \\ 0 & 0 & \varepsilon_z \end{pmatrix}, \tag{4.28}$$

where

$$\varepsilon_a(\omega) = \varepsilon_\infty \left(1 - \frac{\omega_p^2(\omega + i/\tau)}{\omega\left[(\omega + i/\tau)^2 - \omega_c^2 \right]} \right), \tag{4.29}$$

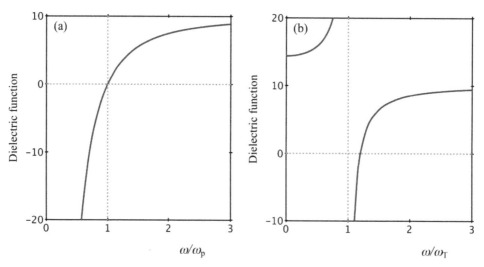

Figure 4.6. Contrasting forms of the scalar dielectric function $\varepsilon(\omega)$ plotted versus frequency for (a) an electron-gas plasma and (b) an ionic crystal, taking spatial dispersion and damping to be negligible in both cases and choosing $\varepsilon_\infty = 10$. In (a) we used equation (1.21) to plot $\varepsilon(\omega)$ versus ω/ω_p, whereas in (b) we used equation (4.35) with $\omega_L/\omega_T = 1.2$ to plot $\varepsilon(\omega)$ versus ω/ω_T.

$$\varepsilon_b(\omega) = \varepsilon_\infty \frac{\omega_p^2 \omega_c}{\omega \left[(\omega + i/\tau)^2 - \omega_c^2 \right]}, \tag{4.30}$$

with ε_z unchanged from the scalar value in equation (4.27). The new angular frequency ω_c is the *cyclotron resonance frequency* (see, e.g., [4]) in a magnetic field of magnitude B_0 and is given by

$$\omega_c = eB_0/m. \tag{4.31}$$

Next we consider the form of the dielectric function in ionic crystals by following the lattice-dynamics approach used by Born and Huang [26]. We consider an infinite diatomic lattice (e.g., as in subsection 2.1.2 for the 1D case). As well as the two sublattices having different masses for the atoms, they are now also associated with opposite charges. It is straightforward to see that it is the long-wavelength ($q \approx 0$) optic phonon that couples to the electromagnetic radiation. This is because a strong electric-dipole moment is created when the two sublattices vibrate in antiphase. Hence if **u** denotes the relative vector displacement for the two sublattices, there will be two contributions to the polarization **P**. One will be proportional to **u** and the other will involve the electronic susceptibility χ and the electric-field vector **E**:

$$\mathbf{P} = \varepsilon_0(\alpha \mathbf{u} + \chi \mathbf{E}), \tag{4.32}$$

where α is a constant. Assuming a time dependence as $\exp(-i\omega t)$, the equation of motion for the displacement gives

$$\left(-\omega^2 - i\omega\Gamma + \omega_T^2 \right)\mathbf{u} = \beta\mathbf{E}_{\mathrm{loc}} = \gamma\mathbf{E}. \tag{4.33}$$

Here the left-hand side contains the usual oscillator terms (with $\Gamma \neq 0$ if there is damping) and ω_T denotes the frequency of the transverse optic phonon. The first term on the right-hand side strictly involves the *local* electric field $\mathbf{E}_{\mathrm{loc}}$ rather than the macroscopic electric field **E**, which is obtained by averaging $\mathbf{E}_{\mathrm{loc}}$ over several unit cells of the crystal. However, it has been argued [24, 26] that these two fields are linearly related and so the final step in equation (4.33) is made (with a new proportionality factor γ).

Using equations (4.32) and (4.33) the polarization can be expressed as

$$\mathbf{P} = \varepsilon_0 \left(\frac{\alpha\gamma}{\omega_T^2 - i\omega\Gamma - \omega^2} + \chi \right)\mathbf{E}. \tag{4.34}$$

It remains only to use equation (1.17) to conclude that the dielectric function, when spatial dispersion is ignored, is

$$\varepsilon(\omega) = \varepsilon_\infty \left(1 + \frac{\omega_L^2 - \omega_T^2}{\omega_T^2 - i\omega\Gamma - \omega^2} \right). \tag{4.35}$$

To achieve this conventional form of the expression, we have defined $\varepsilon_\infty = 1 + \chi$ and

$$\omega_L^2 = \omega_T^2 + \alpha\gamma/\varepsilon_\infty, \tag{4.36}$$

where equation (4.36) defines the longitudinal optic phonon frequency. In the static limit ($\omega \to 0$) of equation (4.35) we obtain

$$\varepsilon(0) = \varepsilon_\infty \omega_L^2 / \omega_T^2, \tag{4.37}$$

which is sometimes known as the Lydane–Sachs–Teller relation. A plot of the frequency dependence predicted by equation (4.35) for an ionic crystal with negligibly small damping is given in figure 4.6(b), where it can be compared with the behaviour in the case of an electron-gas plasma. Specifically, it has been verified that equation (4.35) provides an accurate description of the real and imaginary parts of $\varepsilon(\omega)$ for undoped GaAs [28].

In some materials, such as semiconductors with an appropriate doping level, the plasma frequency and transverse optic phonon frequency might occur in a broadly similar range. In such cases it is appropriate to employ a combined form of equations (4.27) and (4.35) that has both plasma and transverse optic phonon resonances (see [29] for details).

Before concluding this section we briefly consider excitonic materials. These are of interest because their dielectric functions typically exhibit strong spatial dispersion, unlike the previous examples covered. In semiconductors *excitons* can occur as electron–hole pairs in bound states, that arise as a consequence of the electrostatic attraction. Such a bound state may move through a crystal transporting the energy of excitation but not transporting any net charge. Most studies of excitons have focused on the following two limiting cases, e.g., see the book edited by Rashba and Sturge [30]. When the electron–hole attraction is strong (e.g., as in ionic crystals) we have the case of a Frenkel exciton where the electron and hole are close together, effectively on or near the same atom. In contrast, in most semiconductors the Coulomb interaction is strongly screened since the electron and hole are further apart (typically many lattice spacings) and they are only weakly bound to form a Wannier–Mott exciton. It is the latter case that we will explore further.

The Wannier–Mott exciton occurs in a direct-gap, polar semiconductor such as ZnSe or GaAs, so it is of relevance for many photonics and photoluminescence studies. It forms from an electron (effective mass m_e) near the bottom of the conduction band and a hole (effective mass m_h) near the top of the valence band. In a simplified model the excitation energy $\hbar\omega(q)$ of the exciton can be represented as

$$\hbar\omega(q) = E_g - \frac{m^* e^4}{2\hbar^2 \varepsilon_s n^2} + \frac{\hbar^2 q^2}{2M}. \tag{4.38}$$

Here E_g is the band-gap energy and the second term on the right is a hydrogenic-type energy term, in which $m^* = m_e m_h / (m_e + m_h)$ denotes the reduced effective mass, ε_s denotes the static dielectric constant of the semiconductor and $n = 1, 2, \ldots$, is a quantum number corresponding to the states of the system. The third term is an exciton kinetic energy contribution, where $M = m_e + m_h$ is the total exciton mass and $\hbar q$ is the momentum of the centre of mass. Equation (4.38) can be used to estimate the size of the exciton as represented by its effective Bohr radius, which will be large compared with the lattice spacing when ε_s is large and m^* is small.

The exciton couples strongly to light, giving a contribution to the dielectric function which is q-dependent and has the form

$$\varepsilon(q, \omega) = \varepsilon_\infty + \frac{S}{\omega_e^2 + Dq^2 - \omega^2 - i\omega\Gamma}, \qquad (4.39)$$

where ω_e is a shorthand for $\omega(0)$ coming from the first two terms on the right-hand side of the dispersion equation (4.38), $D = \hbar/2M$ and S is an oscillator strength for the exciton. The main difference in form between the above result and the earlier result for the phonon is the strong q-dependent term in the denominator and apart from that the method of derivation via Maxwell's equations is rather similar.

The results for the dielectric functions obtained in this section, along with those for the gyromagnetic susceptibility matrix already found for ferromagnets and antiferromagnets in section 3.3, will be employed in chapter 5 to discuss polaritons in the electromagnetic regime where the excitations couple to photons.

4.4 2D electron gas

Here we examine a limiting case of plasmonics in 2D by taking cases where the mobile charges (electrons or holes) are confined to a sheet of negligible thickness. Such cases can be realized physically at semiconductor heterojunctions, such as the interface between GaAs and $Al_xGa_{1-x}As$ layers, for certain Al concentrations $0 < x < 1$, or as charges trapped at the surface of liquid helium due to the image potential formed there. We will also consider a superlattice formed by a periodic array of charge sheets.

4.4.1 Plasmonics in 2D

We consider a single 2D electron gas (2DEG) localized at the $z = 0$ planar interface, taking without loss of generality a wave propagating in the x-direction with wave vector q_x. The isotropic dielectric media, labelled as A and B, filling the half-spaces above ($z > 0$) and below ($z < 0$), respectively, are characterized by having positive dielectric constants ε_A and ε_B.

The calculation proceeds by considering the electric and magnetic fields in regions A and B subject to the boundary condition at $z = 0$ for the charge sheet. We take the case where the electric field has the form

$$\mathbf{E}_J(\mathbf{r}, t) = (E_{Jx}, 0, E_{Jz})\exp(iq_x x)\exp(iq_{Jz}z)\exp(-i\omega t) \quad (J = A, B), \qquad (4.40)$$

where there is a common q_x and ω, but we need to solve for the components q_{Jz} in the perpendicular direction. In general, they will satisfy

$$q_x^2 + q_{Jz}^2 = \varepsilon_J\omega^2/c^2 \quad (J = A, B). \qquad (4.41)$$

If $q_x^2 \gg \varepsilon_J\omega^2/c^2$, corresponding to the so-called *electrostatic approximation*, then we have $q_{Az}^2 \simeq q_{Bz}^2 \simeq -q_x^2$. The appropriate choice assumed here for localization as $z \to \pm\infty$ in this limit is $q_{Az} = -q_{Bz} = iq_x$ if $q_x > 0$. From Maxwell's equations the corresponding magnetic field must be along the y-direction, i.e., perpendicular to the

x-direction of propagation. This corresponds to a transverse-magnetic (TM) wave, or p-polarization. We note in passing that the other possible choice for the fields, which is the transverse-electric wave or s-polarization, turns out to be uninteresting for mode localization.

The electromagnetic boundary conditions at $z = 0$ require continuity of the tangential electric fields, so $E_{Ax} = E_{Bx}$, while the tangential y-component of the magnetic field are related to the interface charge current and satisfy

$$H_{By} - H_{Ay} = j_x. \qquad (4.42)$$

By analogy with the case of the 3D plasma already discussed in section 4.3 (specifically comparing with equation (4.24) when collisions and spatial dispersion are ignored) we have the linear-response result that

$$j_x = (i\nu e^2/m^*\omega)E_{Ax}, \qquad (4.43)$$

where ν is the number of particles per unit area in the charge sheet and m^* denotes an effective mass. From equations (4.41)–(4.43) and the assumed electric- and magnetic-field polarizations, it is now straightforward to show that the frequency of the localized plasmon in the electrostatic approximation is

$$\omega = \sqrt{\frac{\Omega_0 c q_x}{\varepsilon_A + \varepsilon_B}}, \qquad (4.44)$$

where $\Omega_0 = \nu e^2/\varepsilon_0 mc$ is a characteristic frequency. This result, which was first derived by Stern [31], shows a gapless behaviour at $q_x = 0$ which is quite different from the 3D case, as can be seen on comparing with equation (1.22) at zero wave vector. The inclusion of retardation effects, leading to polaritons, will be described in chapter 5.

4.4.2 Charge-sheets model of superlattices

The analysis just carried out for single charge sheets can be generalized to a periodic array of 2D charge sheets separated by one or more types of dielectric media. This is accomplished by following the transfer-matrix approach as described in section 1.5. The simplest model is to take an infinite array of identical charge sheets with separation D and embedded in the same material (denoted by A) with positive dielectric constant ε_A. This is depicted in figure 4.7, where we show the labelling of the interfaces (and hence the superlattice cells).

The z-axis is chosen to be perpendicular to the interfaces and, as in the previous subsection, we choose the in-plane propagation to be along the x-direction with wave vector q_x. Again we are choosing to consider TM electromagnetic waves, or p-polarization, with $\mathbf{E}$ in the xz-plane and $\mathbf{H}$ in the y-direction. As in the previous subsection we employ the electrostatic approximation, so the allowed values for the analogue of q_{Az} in equation (4.40) are $\pm iq_x$. Therefore within any superlattice cell we need to form solutions for the electric and magnetic-field components that involve a linear combination of terms such as $\exp(q_x z)$ and $\exp(-q_x z)$ that express

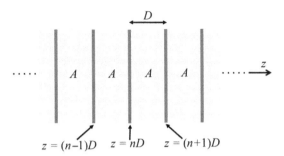

Figure 4.7. Model for an infinite charge-sheet superlattice with periodicity length D. The charge sheets (shown in red) are embedded in a material A.

the z-dependence. For example, the E_x field in the nth cell, taken as the one between the charge sheets at $z = (n - 1)D$ and $z = nD$, can be written as

$$E_x = \exp(iq_x x)\big[a_n \exp[q_x(z - nD)] + b_n \exp[-q_x(z - nD)]\big]\exp(-i\omega t), \quad (4.45)$$

where a_n and b_n are constants. By defining a column vector with the components a_n and b_n we have the basis of a 2×2 transfer-matrix method, following the analysis in section 1.5. We use Bloch's theorem, by analogy with equation (1.58) and also we need to apply the boundary conditions at the interfaces where the charge sheets are located. These boundary conditions are just the same as in the previous subsection, requiring continuity of the tangential electric fields and a discontinuity in the tangential magnetic field as in equation (4.42). The final result for the frequency of the superlattice bulk plasmons is

$$\omega(q_x, Q) = \Omega \left(\frac{q_x D \sinh(q_x D)}{2\big[\cosh(q_x D) - \cos(QD)\big]} \right)^{1/2}, \quad (4.46)$$

where Ω is a characteristic frequency controlled by the areal density ν of charges and the periodic distance D for the superlattice:

$$\Omega = \sqrt{\frac{\nu e^2}{\varepsilon_0 \varepsilon_A m^* D}}. \quad (4.47)$$

The above result for the plasmon modes in the electrostatic (or unretarded) approximation were first obtained by Fetter [32] and Das Sarma and Quinn [33]. The inclusion of retardation is considered in chapter 5 in the context of plasmon-polaritons. Some limiting cases of equation (4.46) can be examined analytically. If $q_x D \gg 1$ it follows that ω is approximately proportional to $\sqrt{q_x}$ in agreement with the result in equation (4.44) for a single embedded charge sheet. On the other hand, if $q_x D \ll 1$ we conclude that, except when $QD = 0$, frequency $\omega \propto q_x$ with a proportionality factor that depends strongly on Q. The general behaviour predicted by equation (4.46) is illustrated in figure 4.8, where the shaded area shows the bulk band and also the lines indicate the frequency versus q_x-dependence for several fixed values of the Bloch wave vector Q.

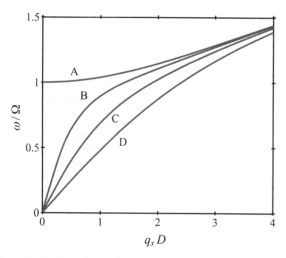

Figure 4.8. Numerical results for the reduced frequency ω/Ω plotted as a function of $q_x D$ for an infinite charge-sheet superlattice with periodicity length D using equation (4.46). The plasmon band is shown by the shaded area and curves 'A', 'B', 'C' and 'D' correspond to QD values of 0, 0.2π, 0.4π and π, respectively.

Several developments are possible from this model. One extension is to the case of a semi-infinite superlattice corresponding to the structure in figure 4.7 but filling only the half-space $z \leqslant 0$. The surface superlattice modes were investigated for this case in [34], where two parameters are found to characterize the behaviour. These are the dielectric constant ε_s of the external medium filling $z > 0$ and the (possibly modified) charge density ν_s for the charge sheet at the external interface $z = 0$. The surface-mode frequency ω then satisfies the equation

$$\mu_s(\mu_s - 1)\sinh(q_x D)(q_x D\omega)^2 + \Omega^2\left(r_s^2 - 1\right)\sinh(q_x D)$$
$$+ \Omega\left[r_s(1 - 2\mu_s)\sinh(q_x D) + \cosh(q_x D)\right] = 0, \tag{4.48}$$

where we denote $r_s = \varepsilon_s/\varepsilon_A$ and $\mu_s = \nu_s/\nu$. Equation (4.48) is applicable provided ω is real and the following condition for localization near the $z = 0$ interface is satisfied:

$$\left|\left[r_s - (\mu_s q_x D\omega/\Omega)\right]\sinh(q_x D) + \cosh(q_x D)\right| < 1. \tag{4.49}$$

Depending on the values of r_s and μ_s, a surface plasmon mode may occur either above or below the bulk plasmon superlattice region. Some numerical examples for these surface modes are given in figure 4.9.

Other extensions of the basic charge-sheets model have included cases where the unit cell is more complicated than in figure 4.7. For example, the charge sheets might be separated by alternating layers A and B of two different materials and/or different thicknesses (see [34]) or the sheets might consist of alternating electron and hole layers [35]. Another application of interest is to n–i–p–i semiconductor superlattices [36], which have four components to the unit cell: an n-type layer, an i-layer (insulating), a p-type layer and another i-layer (with the electron and hole sheets at

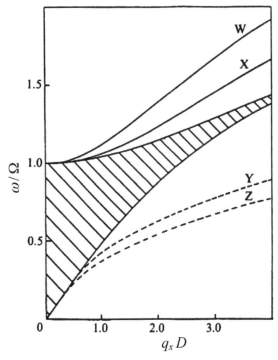

Figure 4.9. Numerical results for ω/Ω plotted as a function of $q_x D$ for an semi-infinite charge-sheet superlattice with periodicity length D using equation (4.48). The bulk band is shown by the shaded area and the curves correspond to surface plasmons: 'W' $r_s = 0.08$ and $\mu_s = 1$; 'X' $r_s = 0.08$ and $\mu_s = 0.75$; 'Y' $r_s = 4$ and $\mu_s = 1$; and 'Z' $r_s = 4$ and $\mu_s = 0.75$. After Constantinou and Cottam [34].

the interfaces of the n- and p-regions, respectively). Various cases of quasi-periodic superlattices, with different growth rules, have been studied [37].

4.5 Bulk-slab model for superlattice plasmons

The superlattice calculations in the previous section are directed to what are sometimes called *intrasubband plasmons*, since it is assumed that the electrons or holes form 2D mobile sheets. It is the case, however, that the charge carriers at a semiconductor interface are trapped by a potential well that has a small spatial width in the z-direction, so the idealization only holds approximately. In practice, there is some degree of delocalization of the surface states and when this is taken into account there may be multiple bound states for the carriers in this potential, leading to more than one subband. The plasmons that occur after allowance for the modifying influence of transitions between these subbands are known as *intersubband plasmons* (see, e.g., [38]).

The bulk-slab model of a superlattice was introduced in section 1.5 where we were mainly concerned with optical properties (photonics) and the dielectric functions of the constituent layers were positive constants. For applications to plasmons it is now assumed that one or more of the constituent layers contains a volume distribution of mobile charges, so the dielectric function takes the form of equation (4.27) in general, or the simpler form given in equation (1.21) if damping and spatial dispersion are ignored.

We start with a two-component $ABABAB...$ superlattice as in section 1.5 and take the non-retarded limit as in the previous section. This allows us to simplify the bulk superlattice dispersion equation (1.64) by taking q_{Az} and q_{Bz} as $\pm i q_x$. Also we assume p-polarization, since this provides the more interesting case experimentally for plasmons, whereupon the required result becomes

$$\cos(QD) = \cosh(q_x d_A)\cosh(q_x d_B) + \frac{1}{2}\left(\frac{\varepsilon_A}{\varepsilon_B} + \frac{\varepsilon_B}{\varepsilon_A}\right)\sinh(q_x d_A)\sinh(q_x d_B). \quad (4.50)$$

A convenient alternative form of this expression [39] is

$$\varepsilon_A/\varepsilon_B = -\beta \pm (\beta^2 - 1)^{1/2} \quad (4.51)$$

with

$$\beta = \frac{\cosh(q_x d_A)\cosh(q_x d_B) - \cos(QD)}{\sinh(q_x d_A)\sinh(q_x d_B)}. \quad (4.52)$$

Then, when ε_A and ε_B both have the bulk plasma form of equation (1.21) with plasma frequencies ω_{pA} and ω_{pB} respectively, we conclude that there are two solutions (denoted by ω^+ and ω^-) for ω. These represent two bulk superlattice bands and their frequencies are given explicitly by

$$\omega^{\pm} = \omega_{pB}\left(1 - \frac{s(1 - r^2)}{s + \beta \mp \left(\beta^2 - 1\right)^{1/2}}\right)^{1/2}. \quad (4.53)$$

We have defined the ratios $r = \omega_{pA}/\omega_{pB}$ and $s = \varepsilon_{\infty A}/\varepsilon_{\infty B}$ in the above.

Much of the experimental work on plasmons in superlattices (see, e.g., [6, 40] for reviews) was based on RS. The semiconductor system GaAs/Al$_x$Ga$_{1-x}$As was the focus of many studies because multilayers with good-quality epitaxial interfaces can be grown for a range of values of the Al concentration x. Typically the multi-quantum-well (MQW) structure is such that the quantum wells (for the electrons) correspond to the GaAs layers, so these will have a volume concentration of electrons while the Al$_x$Ga$_{1-x}$As layers do not. We therefore have a realization of the above theory in which we take medium A as Al$_x$Ga$_{1-x}$As with $\omega_{pA} = 0$ and medium B as GaAs with $\omega_{pB} \neq 0$. Hence $r = 0$ and usually it is a good approximation to take $s = 1$.

The first RS measurements to detect plasmons in GaAs/Al$_x$Ga$_{1-x}$As were reported by Olego et al [41]. They employed a geometry for the experiments in which the incident light was backscattered at 90° from a planar surface. This is the case of $\theta + \phi = 90°$ in the inset (lower left) to figure 4.10, where RS spectra are also shown for three different values of the incident angle θ. The magnitude of in-plane wave vector $q_{\parallel}$ imparted to the plasmon is easily seen to be

$$q_{\parallel} = (2\pi/\lambda)|\sin\theta - \cos\theta| \quad (4.54)$$

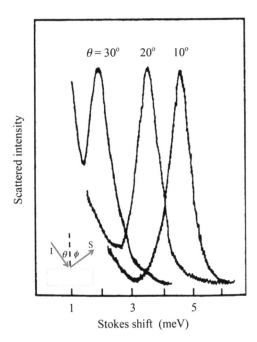

Figure 4.10. RS spectra from plasmons in a superlattice of alternating layers $Al_xGa_{1-x}As$ (medium A) and GaAs (medium B) for three values of the incident angle θ. The layer thicknesses are $d_A \simeq 63$ nm and $d_B \simeq 26$ nm (so $D \simeq 89$ nm). The scattering geometry is illustrated in the inset, where 'I' and 'S' refer to the incident and scattered light, respectively, and $\theta + \phi = 90°$ in the experiments. After Olego *et al* [41].

where λ denotes the vacuum wavelength of the incident light. The three spectra in figure 4.10 clearly illustrate a strong variation in the superlattice plasmon frequency as θ (and hence $q_\parallel$) are changed.

A comparison of the above experiments with theory was carried out by Babiker *et al* [42] and some results are shown in figure 4.11 for the same sample, where the plasmon frequencies are calculated as described above using the bulk-slab model. The full lines ('V', 'W' and 'Y', according to the parameters) correspond to the 'acoustic', or lower-band ω^-, solutions in equation (4.53), whereas the dashed lines ('W', 'Y' and 'Z') correspond to the 'optic', or upper-band ω^+ solutions. The theory lines are characterized in terms of the parameter $\delta \equiv d_B/D$, which is just the fractional thickness of the charge layer. In the limit where $\delta \rightarrow 0$ (case 'V') the results of the bulk-slab model yield the same as the 2DEG charge-sheets model for the ω^- frequency while $\omega^+ \rightarrow \infty$ for the other branch. When $\delta = 0.29$ (case 'W'), corresponding to the experimental data shown in figure 4.10, the ω^- theory curve shifts downwards and is seen to be in good agreement with the RS data. Calculations were also made for the light scattering integrated intensities by means of electric-field Green-function methods.

Finally, we mention that plasmons in quasi-periodic superlattices have been extensively studied, both experimentally (e.g., by RS) and theoretically (using the bulk-slab model and its charge-sheets limit). A review can be found in [37]. In figure 4.12 a comparison is made between calculated and measured plasmon frequencies in a GaAs/$Al_{0.3}Ga_{0.7}As$ Fibonacci superlattice [43]. In this case the building blocks A and B, assembled in sequence as described in subsection 2.4.1 for phonons and in figure 2.13(a),

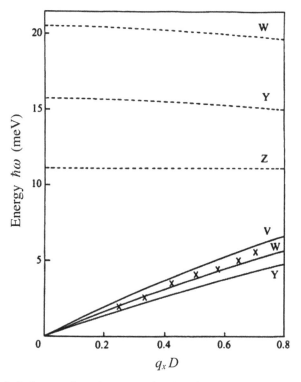

Figure 4.11. Theoretical plasmon dispersion curves for GaAs/Al$_x$Ga$_{1-x}$As superlattices, showing upper (ω^+, dashed lines) and lower (ω^-, full lines) branches for several values of $\delta \equiv d_B/D$: 'V' 0.0; 'W' 0.29; 'Y' 0.6; and 'Z' 1.0. The corresponding experimental ω^- results (crosses) for $\delta = 0.29$ are included [41]. After Babiker *et al* [42].

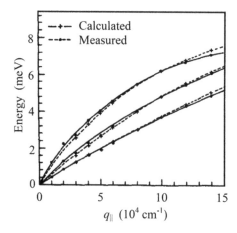

Figure 4.12. Experiment and theory for the lowest three plasmon bands in a GaAs/Al$_{0.3}$Ga$_{0.7}$As Fibonacci superlattice of MQWs. Calculated and measured (by RS) values of energy $\hbar\omega(q_{\parallel})$ at discrete values of in-plane wave vector $q_{\parallel}$. The lines are a guide to the eye. After Merlin *et al* [43].

were *compound* structures. Specifically block *A* consisted of 27 nm of GaAS and 80 nm of $Al_{0.3}Ga_{0.7}As$, while block *B* was 27 nm of GaAS and 43 nm of $Al_{0.3}Ga_{0.7}As$. In the Fibonacci superlattice the $Al_{0.3}Ga_{0.7}As$ barriers were suitably doped so that the MQWs were formed by trapping of the charge carriers in the 27 nm thick GaAs layers, which are unequally spaced due to the growth rule. The theory was carried out using a modified tight-binding approximation for MQWs. It can be seen from the figure that the agreement between experiment and theory for the energy $\hbar\omega(q_{\parallel})$ of the three lowest plasmon bands is excellent.

References

[1] Tamm I 1932 *Phys. Z. Sowjetunion* **1** 733
[2] Goodwin E T 1939 *Proc. Camb. Phil. Soc.* **35** 202 221, 232
[3] Shockley W 1939 *Phys. Rev.* **56** 317
[4] Kittel C 2004 *Introduction to Solid State Physics* 8th edn (New York: Wiley)
[5] Lüth H 1995 *Surfaces and Interfaces of Solid Materials* (Berlin: Springer)
[6] Cottam M G and Tilley D R 2005 *Introduction to Surface and Superlattice Excitations* 2nd edn (Bristol: IOP)
[7] Zangwill A 1988 *Physics at Surfaces* (Cambridge: Cambridge University Press)
[8] Lannoo M and Friedel P 1991 *Atomic and Electronic Structure of Surfaces* (Berlin: Springer)
[9] Northrup J E and Cohen M L 1982 *Phys. Rev. Lett.* **49** 1349
[10] Pavesi L and Lockwood D J (ed) 2004 *Silicon Photonics* (Berlin: Springer)
[11] Novosolev K S, Geim A K, Morozov S V, Jiang D, Zhang Y, Dubonos S V, Gregorieva I V and Firsov A A 2004 *Science* **306** 666
[12] Wallace P R 1947 *Phys. Rev.* **71** 622
[13] Castro Neto A H, Guinea F, Peres N M R, Novosolov K S and Geim A K 2009 *Rev. Mod. Phys.* **81** 109
[14] Power S R and Ferreira M S 2013 *Crystals* **3** 49
[15] Fujita M, Wakabayashi K, Nakada K and Kusakabe K 1996 *J. Phys. Soc. Japan* **65** 1920
[16] Nakada K, Fujita M, Dresselhaus G and Dresselhaus M S 1996 *Phys. Rev.* B **54** 17954
[17] Landau L D and Lifshitz E M 1980 *Statistical Physics* (Oxford: Pergamon)
[18] Cunha A M C, Ahmed M Z, Cottam M G and Costa Filho R N 2014 *J. Appl. Phys.* **116** 013704
[19] Akbari-Sharbaf A and Cottam M G 2014 *J. Appl. Phys.* **116** 194309
[20] Costa-Filho R N, Farias G and Peeters F 2007 *Phys. Rev.* B **76** 193409
[21] Akbari-Sharbaf A, Cottam M G and Fanchini G 2013 *J. Appl. Phys.* **114** 024309
[22] Lahiri J, Lin Y, Bozkurt P, Oleynik I I and Batzill M 2010 *Nat. Nanotechnol.* **5** 326
[23] Zhou J, Zhang S, Wang Q, Kawazoe Y and Jena P 2015 *Nanoscale* **7** 2352
[24] Ashcroft N W and Mermin N D 1976 *Solid State Physics* (New York: Saunders College)
[25] Boardman A D 1982 *Electromagnetic Surface* ed A D Modes (Chichester: Wiley)
[26] Born M and Huang K 1985 *Dynamical Theory of Crystal Lattices* (Oxford: Clarendon)
[27] Lindhard J 1954 *K. Dan. Mat.-Fys. Medd.* **28** 8
[28] Kim O K and Spitzer W G 1979 *J. Appl. Phys.* **50** 4362
[29] Palik E D, Kaplan R, Gammon R W, Kaplan H, Wallis R F and Quinn J J 1976 *Phys. Rev.* B **13** 2497
[30] Rashba E I and Sturge M D (ed) 1982 *Excitons* (Amsterdam: North-Holland)
[31] Stern F 1967 *Phys. Rev. Lett.* **18** 546

[32] Fetter A L 1974 *Ann. Phys. NY* **88** 1

[33] das Sarma S and Quinn J J 1982 *Phys. Rev.* B **25** 7603

[34] Constantinou N C and Cottam M G 1986 *J. Phys.* C **19** 739

[35] Qin G, Giuliani G F and Quinn J J 1983 *Phys. Rev.* B **28** 6144

[36] Farias G A, Auto M M and Albuquerque E L 1988 *Phys. Rev.* B **38** 12540

[37] Albuquerque E L and Cottam M G 2003 *Phys. Rep.* **376** 225

[38] Eliasson G, Hawrylak P and Quinn J J 1987 *Phys. Rev.* B **35** 5569

[39] Camley R E and Mills D L 1984 *Phys. Rev.* B **29** 1695

[40] Pinczuk A 1984 *J. Phys.* **45** C5–477

[41] Olego D, Pinczuk A, Gossard A C and Wiegmann W 1982 *Phys. Rev.* B **25** 7867

[42] Babiker M, Constantinou N C and Cottam M G 1986 *Solid State Commun.* **59** 751

[43] Merlin R, Valladares J P, Pinczuk A, Gossard A C and English J H 1992 *Solid State Commun.* **84** 87

IOP Publishing

Dynamical Properties in Nanostructured and Low-Dimensional Materials

Michael G Cottam

Chapter 5

Polaritons

The basic concept of polaritons was introduced in chapter 1 in terms of the coupling of an electromagnetic wave (a photon) to a crystal excitation in an unbounded homogeneous dielectric medium. For example, simple results for a bulk plasmon-polariton were presented in equation (1.22) and figure 1.3 for unbounded media. In this chapter we discuss various different kinds of polaritons, both non-magnetic and magnetic, at surfaces and interfaces and in nanostructured materials. The role of the dielectric function of the constituent materials (or the magnetic susceptibility function for magnetic materials) will be important and we will make use of results derived for those quantities in chapters 3 and 4. Some excellent accounts of polaritons can be found in the volumes edited by Burstein and de Martini [1], Agranovich and Mills [2], and Boardman [3], while the more recent book by Albuquerque and Cottam [4] focuses on multilayer structures that are either periodic or quasi-periodic. Related experimental techniques, including the method of ATR and other optical techniques, are reviewed in [5, 6].

5.1 Phonon-polaritons

We start with the well-known result found in most solid-state physics textbooks (see [7]) for the bulk phonon-polaritons in an ionic crystal. This is obtained directly by putting the damping equal to zero in equation (4.35) for the dielectric function, which then simplifies with the aid of equations (4.36) and (4.37) to give

$$\varepsilon(\omega) = \varepsilon_\infty \left(\frac{\omega_L^2 - \omega^2}{\omega_T^2 - \omega^2} \right). \tag{5.1}$$

Next we substitute $\varepsilon(\omega)$ into equation (1.22) to obtain the dispersion relation. The resulting quadratic equation in ω^2 yields two solutions, corresponding to a dispersion relation with two branches, as shown in figure 5.1. The light line

doi:10.1088/978-0-7503-1054-3ch5

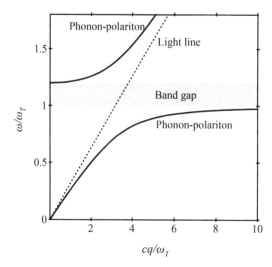

Figure 5.1. Dispersion relation for a linear phonon-polariton (solid lines) in a 3D ionic material, as given by equations (1.22) and (5.1). The frequency, in terms of dimensionless ω/ω_T is plotted versus dimensionless wavenumber, in terms of cq/ω_T taking $\omega_L/\omega_T = 1.2$ and $\varepsilon_\infty = 10$.

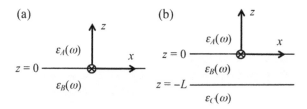

Figure 5.2. Assumed geometry and notation for polariton calculations in non-magnetic materials. (a) Case of a single interface (at $z = 0$) between two dielectric media A and B. (b) Case of two interfaces (at $z = 0$ and $z = -L$) involving dielectric media A, B and C.

corresponds to $\omega = cq/\sqrt{\varepsilon_\infty}$, which represents one of the asymptotic solutions as $q \to \infty$, the other being $\omega = \omega_T$. There is a stop band (or band gap) for the range $\omega_T < \omega < \omega_L$; this is where $q^2 < 0$ and hence where we might expect to find localized modes when surfaces are introduced.

5.1.1 One-interface geometry

We suppose now that there is a single planar interface at $z = 0$ between two semi-infinite dielectric media described by isotropic dielectric functions $\varepsilon_A(\omega)$ and $\varepsilon_B(\omega)$ as illustrated in figure 5.2. The mode propagation parallel to the interface will be taken along the x-axis with wave vector denoted by q_x. It can be verified that there is no surface mode corresponding to the electric field in the y-direction and therefore we restrict attention here to the case where the electric field $\mathbf{E}_J(\mathbf{r}, t)$ in each of the media $J = A, B$ is in the xz-plane, written as in equation (4.40) for a p-polarized or TM wave. Also we must have $\text{Im}(q_{Az}) > 0$ and $\text{Im}(q_{Bz}) < 0$ for localization, where both quantities satisfy equation (4.41). If the ε_J are real, it follows that $q_x^2 > \varepsilon_J \omega^2/c^2$ for localization. In fact, we may write $q_{Az} = i\kappa_A$ and $q_{Bz} = -i\kappa_B$ where

$$\kappa_J = \sqrt{q_x^2 - \varepsilon_J\left(\omega^2/c^2\right)} > 0 \quad (J = A, B). \tag{5.2}$$

One of Maxwell's equations to be satisfied at the $z = 0$ interface is $\nabla \cdot \mathbf{D} = 0$, since there is no charge sheet in the present calculation. This implies for the electric-field components that

$$q_x E_{Jx} + q_{Jz} E_{zJ} = 0 \quad (J = A, B). \tag{5.3}$$

Other boundary conditions require the continuity of the tangential $\mathbf{E}$ field and the normal $\mathbf{D}$ field, giving

$$E_{Ax} = E_{Bx}, \quad \varepsilon_A E_{Az} = \varepsilon_B E_{Bz}. \tag{5.4}$$

The four equations in (5.3) and (5.4) constitute a solvability condition for the four electric-field components leading to the condition that

$$q_{Az}/\varepsilon_A = q_{Bz}/\varepsilon_B, \tag{5.5}$$

and substitution into equation (4.41) for q_x^2 gives the surface polariton dispersion relation in the form

$$q_x^2 = \frac{\omega^2}{c^2}\left(\frac{\varepsilon_A \varepsilon_B}{\varepsilon_A + \varepsilon_B}\right). \tag{5.6}$$

We have already established that q_{Az} and q_{Bz} have opposite signs, so it follows from equation (5.5) that this must also be true for ε_A and ε_B at the surface-mode frequency. We therefore conclude that two necessary conditions are $\varepsilon_A \varepsilon_B < 0$ and $\varepsilon_A + \varepsilon_B < 0$, where the latter condition is deduced from equation (5.6). These requirements can be satisfied in various ways and in many experimental situations it may be that one of the media (say A) has a positive (and often frequency-independent) dielectric constant ε_A, while the other medium B (called the *surface-active medium*) has the negative and frequency-dependent dielectric function.

So far, our discussion of surface polaritons has been quite general, but now we particularize to the phonon-polariton case. Specifically we choose medium A to be a vacuum or air, so $\varepsilon_A = 1$, and medium B to be an ionic crystal with real dielectric function given by equation (5.1). It follows from our discussion in the previous paragraphs that a surface phonon-polariton can exist in the frequency range corresponding to $-\infty < \varepsilon(\omega) < -1$, which means explicitly $\omega_T < \omega < \omega_S$ where

$$\omega_S = \sqrt{\frac{\varepsilon_\infty \omega_L^2 + \omega_T^2}{\varepsilon_\infty + 1}}. \tag{5.7}$$

This characteristic frequency ω_S lies in the band gap between ω_T and ω_L. The predicted dispersion of the surface phonon-polariton is illustrated in figure 5.3 for

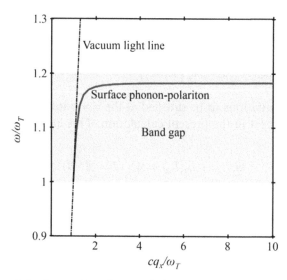

Figure 5.3. Dispersion relation for a surface phonon-polariton (solid line) localized near the planar interface between a vacuum and an isotropic ionic material, assumed to be described by the same parameters as in figure 5.1. The vacuum light line corresponding to $\omega = cq_x$ is also shown.

the same choice of active-medium parameters as previously for the bulk case in figure 5.1. We note that the surface polariton exists only to the right of the *vacuum light line*, from which it emerges at $\omega = \omega_T$. The explicit algebraic form of the ω versus q_x relationship for the surface mode is quoted, for example, in [4].

We note that the results derived here become much more complicated when one or both of the media are anisotropic. Discussions of some of the cases that arise under these circumstances can be found, for example, in [2, 4].

5.1.2 Two-interface geometries

Next we turn to the case of two-interface modes by considering the geometry depicted in figure 5.2(b). Medium B is now a slab of finite thickness L, while it is bounded on either side by semi-infinite media labelled A and C. The analysis can be carried out essentially as described for polaritons in the previous single-interface geometry, except that we now write down solutions for the electric-field vector in all three regions and the analogous electromagnetic boundary conditions are applied at the $z = 0$ and $z = -L$ interfaces. Many different cases can arise, depending on factors such as the choice for polarizations of the modes, which of the media are surface active (i.e., have an explicitly ω-dependent dielectric function), whether the media are isotropic or otherwise, whether damping can be ignored and so forth. Results for most of these cases can be found in the very extensive literature now available; some reviews with many further references are provided in [2, 4, 8]. Some particular examples of interest are highlighted below.

When all three media are isotropic and described by different real, scalar $\varepsilon_J(\omega)$ (with $J = A$, B, C), the type of calculation described above leads to the implicit dispersion relation for the surface polaritons in the form

$$\left(\frac{\varepsilon_A(\omega)q_{Bz} + \varepsilon_B(\omega)q_{Az}}{\varepsilon_A(\omega)q_{Bz} - \varepsilon_B(\omega)q_{Az}}\right)\exp(-2iq_{Bz}L) = \left(\frac{\varepsilon_C(\omega)q_{Bz} + \varepsilon_B(\omega)q_{Cz}}{\varepsilon_C(\omega)q_{Bz} - \varepsilon_B(\omega)q_{Cz}}\right). \tag{5.8}$$

The complex quantities q_{Jz} are defined analogously to equation (4.41), but the localization criteria are modified for the three-layer structure. We require $\text{Im}(q_{Az}) > 0$ and $\text{Im}(q_{Cz}) < 0$, leading to

$$q_x^2 > \left(\omega^2/c^2\right)\max(\varepsilon_A, \varepsilon_C). \tag{5.9}$$

On the other hand, there is no such requirement involving $\text{Im}(q_{Bz})$. Assuming $\varepsilon_B(\omega)$ to be real (i.e., negligible damping in the film), q_{Bz} can be either real or imaginary according to whether q_x^2 is less than or greater than $(\omega^2/c^2)\varepsilon_B$, respectively. If q_{Bz} is imaginary we have a decay-like behaviour away from both interfaces, which can be interpreted as the surface polariton at one interface being perturbed by the surface polariton at the other interface. The two surface modes overlap in their amplitude terms and become shifted in frequency giving upper- and lower-frequency surface branches. In the other situation, where q_{Bz} is real, there is a wave-like behaviour for the z-dependence of the electric-field components within the film. This corresponds to a *guided-wave polariton*, where the light wave is effectively trapped in the film by total internal reflections at the interfaces.

The general dispersion relation given by equation (5.8) is complicated, but there is a useful simplification for a symmetric geometry in which the bounding media A and C are the same and are dispersionless, i.e., $\varepsilon_A(\omega) = \varepsilon_B(\omega) \equiv \varepsilon_A$. Localization requires $q_{Az} = -q_{Cz} = i\kappa_A$, with κ_A defined as in equation (5.2), whereupon equation (5.8) can be rearranged and factorized to give two solutions. For the surface-polariton case (when imaginary $q_{zB} \equiv i\kappa_B$) the results are

$$\begin{aligned} \varepsilon_B(\omega) &= -(\varepsilon_A\kappa_B/\kappa_A)\tanh(\kappa_B L/2), \\ \varepsilon_B(\omega) &= -(\varepsilon_A\kappa_B/\kappa_A)\coth(\kappa_B L/2). \end{aligned} \tag{5.10}$$

The hyperbolic functions in the above equations both tend to unity as the film thickness $L \to \infty$ and so we recover the dispersion relations for single-interface modes by analogy with equation (5.5). For the p-polarized guided-wave polaritons where q_{Bz} is real in this symmetric geometry the results become

$$\begin{aligned} \varepsilon_B(\omega) &= -(\varepsilon_A q_{Bz}/\kappa_A)\tan(q_{Bz}L/2), \\ \varepsilon_B(\omega) &= -(\varepsilon_A q_{Bz}/\kappa_A)\cot(q_{Bz}L/2). \end{aligned} \tag{5.11}$$

Numerical results deduced for the surface phonon-polariton frequencies at finite L are shown in figure 5.4 using equation (5.1) for $\varepsilon_B(\omega)$ and parameters appropriate for LiF films bounded by different materials [9]. Various cases that emphasize the distinction between the upper and lower modes are included, taking examples for both symmetric and asymmetric geometries through choices of different substrate materials. The two modes of an unsupported film (with media A and C being vacuums) correspond to the curves labelled 'a' and 'b' which show quite different behaviour. Curve 'c' applies for an infinitely thick film, irrespective of the substrate

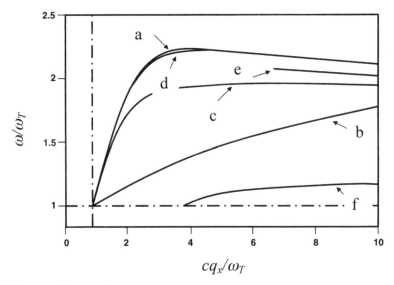

Figure 5.4. Upper- and lower-surface phonon-polaritons for a LiF film bounded by various media. The curves are as follows: 'a' upper branch for an unsupported film with $\omega_T L/c = 0.1$; 'b' lower branch for an unsupported film with $\omega_T L/c = 0.1$; 'c' upper branch for $L \to \infty$ on any substrate; 'd' upper branch for a film with $\omega_T L/c = 0.05$ on a K substrate (with $\omega_p = 3.8$ eV); 'e' upper branch for a film with $\omega_T L/c = 0.1$ on a Si substrate (with $\varepsilon_C = 11.7$); and 'f' lower branch for a film with $\omega_T L/c = 0.1$ on a Si substrate. After Oliveira et al [9].

material. Curve 'd' illustrates the minimal difference made (compared to case 'a') when the lower half of the film is replaced by a metal substrate. Curves 'e' and 'f' refer to an asymmetric geometry where the film is deposited on a Si substrate.

It has been assumed so far that damping effects are negligible in all the media and so the dielectric functions are real. The extensions of the theory to include complex dielectric functions in the two-interface geometry were made first by Fukui et al [10] and Sarid [11] in the context of plasmon-polaritons. In particular, they identified a mode referred to as the *long-range surface plasmon* (LRSP), which will be mentioned in section 5.2 on plasmon-polaritons. Analogous investigations of the role of damping in the phonon-polariton case are described in [12].

The observation of surface polaritons has required the development of specialized techniques. This is because, in a planar geometry as discussed so far, the q_x and ω values satisfy the inequality $q_x^2 > \varepsilon_B \omega^2/c^2$ in a medium B. On the other hand, a conventionally propagating light beam incident in the medium satisfies $q_x^2 + q_z^2 = \varepsilon_B \omega^2/c^2$ with q_z real. This means that the surface polariton cannot be excited directly by the light beam and is therefore referred to as a *non-radiative* mode. The technique of ATR was developed to overcome this limitation, essentially by removing the restriction on q_z to be real. The usual configuration, as introduced by Otto [5], is sketched in figure 5.5. It involves a coupling prism with real positive dielectric constant ε_P on top of a spacer medium (often air or a vacuum) to form a gap, with the surface-active medium (the sample being studied) as the third medium. The prism is chosen to have a relatively large refractive index, so that $\varepsilon_P > \varepsilon_A$. Then if

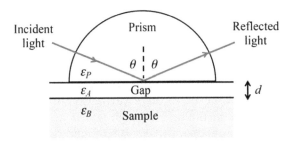

Figure 5.5. Schematic geometry for an ATR experiment in the Otto configuration. The surface-active medium is shown shaded.

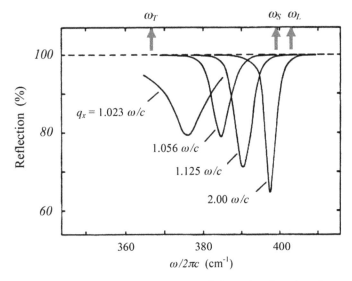

Figure 5.6. ATR spectra for the intensity reflection coefficient plotted against frequency for surface phonon-polaritons in GaP. Each curve is labelled with its q_x value as the angle θ is varied. After Marschall and Fischer [13].

the direction of the incident light is chosen such that angle θ exceeds the critical angle θ_c for total internal reflection at the prism–gap interface, where

$$\theta_c = \sin^{-1}\left(\sqrt{\varepsilon_A/\varepsilon_P}\right), \tag{5.12}$$

there will be an evanescent wave in the gap (decaying with respect to the z-direction as previously defined). This can excite a surface polariton in the active layer B (the sample), which will be observed as a dip in the measured intensity reflection coefficient when ω and $q_x = \sqrt{\varepsilon_P}(\omega/c)\sin\theta$ both lie on the surface polariton dispersion curve. It is straightforward, based on the analysis earlier in this subsection, to calculate an expression for the dependence of the reflection coefficient on frequency (see, e.g., [8]).

An example showing several ATR spectra at different q_x values is given in figure 5.6. The dips correspond to surface phonon-polaritons excited in GaP [13]. In accordance with the theory presented earlier, the frequencies where the reflection is a minimum lie between ω_T and ω_S, where the latter quantity is defined in equation (5.7).

Variations of the standard Otto ATR method include the Raether–Kretschmann ATR method, in which the surface-active medium is grown by evaporation directly on to the prism (i.e., there is no gap), and the grating method, in which coupling to the non-radiative mode is achieved through having a diffraction grating laid down on the surface of the active medium. Thorough review accounts of these methods, together with examples of applications, are to be found in [6, 14].

As mentioned earlier, RS is a suitable technique in general for observing the phonon-polaritons in the geometries under discussion. With surface phonon-polaritons, however, the conditions for the scattering cross section to be sufficiently large for observations are more stringent. The usual backscattering configuration is unfavourable, but it was demonstrated by Evans *et al* [15] that the near-forward RS by transmission through a film leads to a reinforced scattered signal.

5.1.3 Multiple-interface geometries

For the case of a periodic superlattice we have already derived a general result for the frequencies of the optical modes in subsection 1.5.1. There we considered a two-component *...ABABAB...* structure with alternating layers as depicted in figure 1.8. Although we had in mind the photonics properties of the structure, for which ε_A and ε_B in this bulk-slab model are usually dispersionless (i.e., positive constants), the formal results are quite general and include cases where either (or both) dielectric functions have spatial dispersion as in equation (5.1). Therefore in the present context, equations (1.64)–(1.66) for the bulk modes of the superlattice having either the s- or p-polarization are still applicable.

The analytic expressions become quite formidable, even when just one of the media (say B) has spatial dispersion with $\varepsilon_B(\omega)$ given by equation (5.1) while ε_A is a constant independent of ω. Consequently, approximations are frequently used, depending on the superlattice periodic length D and the wavelengths of the superlattice modes that are relevant in any experiment.

One useful simplification is known as the *effective medium approximation* and broadly it applies when D is large compared with the interatomic spacings in the media A and B but small compared with any of the relevant optical wavelengths. In practice, for our assumed geometry and notation, the latter requirement means that $QD \ll 1$, $q_x D \ll 1$, $q_{Az} D \ll 1$ and $q_{Bz} D \ll 1$. We may then expand the trigonometric terms in the Rytov equation (1.64) for small values of their arguments, neglecting terms that are cubic and higher order in small quantities. The result for the dispersion relation, as found by Raj and Tilley [16], depends on the polarization and takes the form

$$\frac{Q^2 + q_x^2}{\bar{\varepsilon}_{xx}} = \frac{\omega^2}{c^2} \tag{5.13}$$

for s-polarization and

$$\frac{Q^2}{\bar{\varepsilon}_{xx}} + \frac{q_x^2}{\bar{\varepsilon}_{zz}} = \frac{\omega^2}{c^2} \tag{5.14}$$

for p-polarization. Here $\bar{\varepsilon}_{xx}$ and $\bar{\varepsilon}_{zz}$ are tensor components of an effective dielectric function defined by

$$\bar{\varepsilon}_{xx} = \frac{\varepsilon_A d_A + \varepsilon_B d_B}{D},$$

$$\frac{D}{\bar{\varepsilon}_{zz}} = \frac{d_A}{\varepsilon_A} + \frac{d_B}{\varepsilon_B}. \tag{5.15}$$

The above result shows that the superlattice can be represented by an effective medium with an anisotropic dielectric function.

As discussed in [16], at the long wavelengths where the above approach is valid the electromagnetic boundary conditions that E_x and D_z are continuous at any interface mean that these field components have essentially the same values (denoted by $\bar{E}_x$ and $\bar{D}_z$, respectively) in either the A or B layers over many periods of the superlattice. The other field components are found from

$$D_{Jx} = \varepsilon_0 \varepsilon_J E_{Jx}, \qquad E_{Jz} = D_{Jz}/\varepsilon_0 \varepsilon_J, \quad (J = A, B). \tag{5.16}$$

It may then be verified that, on defining $\bar{D}_x$ and $\bar{E}_z$ as the spatial averages of these field components over one superlattice period, we have the effective-medium relationships

$$\bar{D}_x = \varepsilon_0 \bar{\varepsilon}_{xx} \bar{E}_x, \qquad \bar{D}_z = \varepsilon_0 \bar{\varepsilon}_{zz} \bar{E}_z, \tag{5.17}$$

where the tensor components of the effective dielectric function are those found in equation (5.15). Other formulations of the effective medium approximation for layered superlattices are given in [17, 18]. Also the method has been extended to surface modes of superlattices (see, e.g., [16, 19]).

The periodic layered superlattice is just one example of a *composite medium*. Other examples occur if we have periodic arrays of spheres or cylindrical wires embedded in a matrix of another material, again leading to effective-medium expressions for components of the dielectric function. A discussion is given by Bergman and Stroud [20].

An example of some experimental data on superlattice phonon-polaritons, along with its successful interpretation using the effective medium approximation, is given in figure 5.7. This shows normal-incidence far-infrared reflectivity on a GaAs/Al$_x$Ga$_{1-x}$As superlattice [21]. The measurements were carried out using dispersive Fourier-transform spectroscopy which provides simultaneous data for the amplitude and phase of the reflectivity. The theory (full lines) was obtained within the effective-medium approximation described above and gives good agreement with the data. The superlattice was composed of alternating layers of GaAs with thickness 5.5 nm and Al$_x$Ga$_{1-x}$As with thickness 17 nm.

Apart from the effective-medium approach, the other approximation for the general formalism of polaritons in superlattices with spatially dispersive media involves short-period superlattices. Specifically it refers to the properties of *confined modes* in these structures, which occur if one of the quantities q_{Az} and q_{Bz} is real,

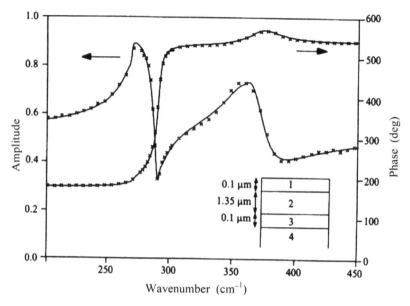

Figure 5.7. Normal-incidence far-infrared reflectivity on a GaAs/Al$_x$Ga$_{1-x}$As superlattice, showing both the amplitude (left scale) and phase angle (right scale). The sample geometry is illustrated in the inset, where the superlattice (region 2) is sandwiched between Al$_{0.35}$Ga$_{0.65}$As (regions 1 and 3) on top of a GaAs substrate (region 4). The crosses and lines correspond to experimental data and theory, respectively. After Maslin *et al* [21].

while the other is not. Suppose it is q_{Bz} that is real; from equation (4.41) this means that

$$\varepsilon_B > c^2 q_x^2 / \omega^2 > \varepsilon_A. \tag{5.18}$$

If thickness d_B is small (i.e., comparable with the optical wavelengths involved) the confined modes in this case have very nearly the character of standing waves across medium B. If the wave-vector quantization is such that an integer number of half-wavelengths fit into the thickness d_B then

$$q_{Bz} = n\pi / d_B, \quad n = 1, 2, \ldots n_B, \tag{5.19}$$

where $n_B = d_B/a$ ($a =$ lattice constant) is the number lattice units in layer B. The mode frequencies are then given to a good approximation by inserting the above quantized q_{Bz} values into the dispersion relation for bulk phonon-polaritons in medium B. In practice, a better fit to the experimental data can be obtained if d_B in equation (5.19) is replaced by $d_B^* = (n_B + \delta)a$ to allow for a small amount of penetration of the phonon-polariton mode into medium A (see, e.g., [22]). Often the value $\delta \simeq 0.8$ is adopted for the dimensionless δ.

5.2 Plasmon-polaritons

We recall that in chapter 1 the example of a bulk electron plasma was used to introduce the concept of a polariton. There we employed equation (1.21) for $\varepsilon(\omega)$ of a plasma, in the absence of damping and spatial dispersion, to obtain the bulk plasmon-polariton dispersion relation quoted in equation (1.22) and depicted in figure 1.3. To summarize, it consists of a single branch with a band gap extending in frequency from 0 to ω_p and its limiting forms for $cq/\omega_p \ll 1$ and $cq/\omega_p \gg 1$ are already given.

By analogy with the phonon-polariton case in the previous section of this chapter we anticipate that the surface plasmon-polariton associated with interfaces will occur within the above band gap. As in the case where plasmons at surfaces and interfaces were described in chapter 4, there are broadly two models that can be followed for the plasmon-polaritons depending on the degree of electron localization. These are the charge-sheets (or 2DEG) model and the bulk-slab model, where we now generalize the formalism to include the electromagnetic retardation effects, previously neglected in sections 4.4 and 4.5.

5.2.1 2D electron gas model

We start with the same single-interface model as in subsection 4.4.1 which consisted of a 2DEG at the $z = 0$ planar interface between dielectric media A and B filling the half-spaces $z > 0$ and $z < 0$, respectively. Equations (4.40)–(4.43) are still applicable when we consider the case of p-polarization (with $\mathbf{E}$ in the xz-plane and $\mathbf{H}$ in the y-direction).

The difference compared with chapter 4 is that we now retain the full form of q_{Jz} (with $J = A, B$), written as $q_{Az} = i\kappa_A$ and $q_{Bz} = -i\kappa_B$ with κ_A and κ_B both real and positive and defined as in equation (5.2). By following the same steps as before, we find that equation (4.44) for the dispersion in the electrostatic case is replaced by the implicit dispersion relation

$$\frac{\varepsilon_A}{\kappa_A} + \frac{\varepsilon_B}{\kappa_B} = \frac{\Omega_0 c}{\omega^2}, \tag{5.20}$$

where $\Omega_0 = \nu e^2/\varepsilon_0 mc$ as before. It is worth noting that equation (5.20) gives rise to a localized polariton mode even when the bounding media A and B are identical. The general result given above can be rearranged in this special case as

$$8\varepsilon_A\omega^2 \big/ \Omega_0^2 = \left[1 + \left(4cq_x/\Omega_0\right)^2\right]^{1/2} - 1. \tag{5.21}$$

This expression describes a single plasmon-polariton branch with its angular frequency ω increasing monotonically from zero as q_x is increased in magnitude. For small q_x, such that $cq_x \ll \Omega_0$, the predicted behaviour is that $\omega \propto q_x$. The dependence becomes $\omega \propto \sqrt{q_x}$ at larger q_x (such that $cq_x \gg \Omega_0$).

Next for plasmon-polaritons in the charge-sheets model of a superlattice we may straightforwardly generalize results in subsection 4.4.2. The model and geometry for the superlattice are the same as in figure 4.7 and the method of calculation is similar, but the results with retardation are more complicated. Instead of the explicit result in equation (4.46) the bulk-mode frequency is given implicitly by the following generalization:

$$\cos(QD) = \cos(q_{Az}D) + \frac{\Omega^2 q_{Az}D}{2\omega^2} \sin(q_{Az}D), \tag{5.22}$$

where Ω is the characteristic frequency related to the areal charge density ν and the periodic length D as specified in equation (4.47). The wave-vector component q_{Az} in medium A is

$$q_{Az} = \left[\left(\varepsilon_A \omega^2/c^2\right) - q_x^2\right]^{1/2}. \tag{5.23}$$

This quantity can be imaginary (as in the limit of negligible retardation) or real depending on the value of wave vector q_x.

Just as described for the unretarded plasmons in charge-sheet superlattices (see subsection 4.4.2) the above simple model can be extended in various ways [23–25]. The extensions include surface modes in semi-infinite or finite superlattices, having alternating A and B layers of different media between the charge (electron or hole) sheets, and applications to n–i–p–i semiconductor superlattices. A thorough review account, also covering quasi-periodic structures, is given in [4].

5.2.2 Bulk-slab model

For this case there are no charge sheets (2DEG) at the interfaces and the charges are distributed in the volume of one or more dielectric media. In the single-interface geometry we assume the same geometry as in figure 5.1(a), leading once again to equation (5.6) as giving the formal expression for the surface-polariton dispersion relation, provided the necessary conditions $\varepsilon_A \varepsilon_B < 0$ and $\varepsilon_A + \varepsilon_B < 0$ can be satisfied.

As a particular example, we may take medium A as vacuum or air, so $\varepsilon_A = 1$, while medium B is an electron plasma as in a metal or doped semiconductor with its real dielectric function given by equation (1.21). We then conclude that the surface plasmon-polariton exists in the frequency interval $0 < \omega < \omega_S$ where

$$\omega_S = \omega_p \sqrt{\frac{\varepsilon_\infty}{\varepsilon_\infty + 1}}, \tag{5.24}$$

and the dispersion relation of the mode is easily found. We recall that the bulk plasmon-polaritons in medium B satisfy the dispersion relation in equation (1.22), so they occur for $\omega > \omega_p$ as shown in figure 1.3. The surface and bulk polariton frequencies are now plotted in figure 5.8 in terms of cq_x/ω_p, where we assume $\varepsilon_\infty = 1$ so $\omega_S/\omega_p = 1/\sqrt{2} \simeq 0.71$.

Generalizing now to a superlattice, the formal expressions for the dispersion relations are essentially the same as already described for phonon-polaritons in subsection 5.1.3 on the basis of the bulk-slab model. The only essential difference is that the plasma-medium form for the dielectric function, either the simplified form in equation (1.22) or the more general expression in (4.27), needs to be employed instead of the previous dielectric function for an ionic solid. Likewise the simplifications that were discussed in subsection 5.1.3 using the effective-medium approximation and/or the treatment of confined modes are also applicable for plasmon-polaritons [4].

Following advances in the techniques for the growth of semiconductor materials in layers with abrupt interfaces (see, e.g., [26]), it is now feasible to produce

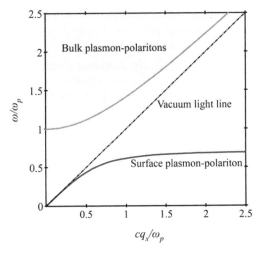

Figure 5.8. Dispersion relation for a surface plasmon-polariton (solid line) localized near the planar interface between vacuum and a plasma material (such as n-doped GaAs). The vacuum light line corresponding to $\omega = cq_x$ is shown, as well as the continuum region of bulk plasmon-polaritons.

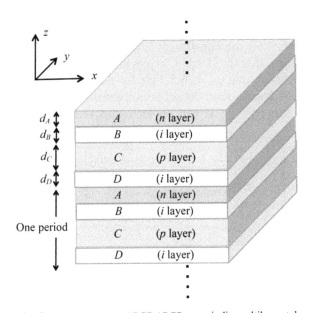

Figure 5.9. Structure of a four-component $...ABCDABCD...$ periodic multilayer taken as a model for an n–i–p–i semiconductor superlattice.

semiconductor superlattices or arrays with a more complicated growth pattern than the periodic $...ABABAB...$ structure. One notable example regarding the study of plasmon-polaritons (either retarded or unretarded) is the case of periodic n–i–p–i superlattices, which were mentioned briefly in section 4.4. They involve a repeating pattern with four layers as illustrated in figure 5.9: A, an n-type semiconductor with a volume distribution of electron carriers; B, an insulating ('i') medium; C, a p-type semiconductor with a volume distribution of hole carriers; and D, another 'i' medium. These layers might consist of, for example, layers of n-doped and p-doped GaAs together with intrinsic (pure) GaAs. The doped layers are crystallographically only slightly perturbed by the relatively low concentration of dopants, typically 10^{17} to 10^{19} cm^{-3}. The layers might be of any arbitrary thickness, as indicated, and the periodicity length is $D = d_A + d_B + d_C + d_D$. If the thicknesses d_A and d_C of the charge layers are small enough, they might be replaced by 2D charge sheets (2DEG or its hole gas counterpart for the n- and p-type layers, respectively) as an approximation. Due to their more complicated structure and the novel features expected for the plasmon-polaritons, n–i–p–i superlattices have been extensively investigated since the late 1980s (see [25, 27]).

The plasmon-polariton modes can be studied by a straightforward extension of the transfer-matrix method described in subsection 1.5.1. Briefly, a 2×2 matrix representation can be defined as before in terms of forward- and backward-travelling waves for components of the electric field and transformation relations across each layer can be formed analogously to equations (1.50)–(1.56). The resulting expression for the transfer matrix T involves a product of more terms than in equation (1.57) because there are now four layers in the repeating periodic arrangement [4, 25, 27].

An example of a calculation for the dependence of the plasmon-polariton frequencies on the in-plane wave vector q_x is given in figure 5.10. It has been assume here [25] that all the constituent layers have the same thickness equal to 20 nm (so $D = 80$ nm), that the A and C layers are Si doped with n- and p-impurities to the same extent (such that $\omega_p = 76.5$ THz for both layers) and that the B and D layers are SiO$_2$. The background dielectric constants ε_∞ are taken as 11.7 and 12.3 for Si and SiO$_2$, respectively. It can be seen that there are four bulk bands, as

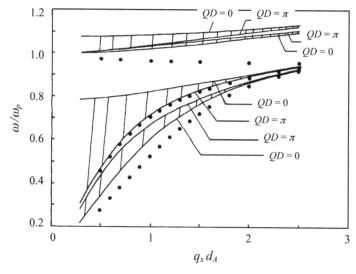

Figure 5.10. Frequencies of plasmon-polaritons plotted versus dimensionless in-plane wave vector $q_x d_A$ for an n–i–p–i semiconductor superlattice, showing four bulk modes (shaded areas) and three surface branches (dots). See the text for parameter values. After Farias *et al* [25].

expected for a four-component superlattice. These are shown shaded and have boundaries corresponding to $QD = 0$ and $QD = \pi$. For a semi-infinite n–i–p–i superlattice in this case it is found that there are three surface modes (shown by dotted lines).

5.3 Magnon-polaritons

The magnons or SWs were discussed in chapter 3 in various regimes of the wave vector, but we have left aside until now the electromagnetic regime in the dipole-dominated case (see figure 3.1). Many aspects of the treatment for non-magnetic polaritons will be applicable here, but with the role of the dielectric function being replaced by the permeability tensor which has a gyromagnetic form [28]:

$$\bar{\mu} = \begin{pmatrix} \mu_{xx} & \mu_{xy} & 0 \\ -\mu_{xy} & \mu_{xx} & 0 \\ 0 & 0 & \mu_{zz} \end{pmatrix}, \tag{5.25}$$

where $\mu_{xx} = \mu_{\infty}(1 + \chi_a)$, $\mu_{xy} = i\mu_{\infty}\chi_b$ and $\mu_{zz} = \mu_{\infty}$. The magnetic susceptibility terms χ_a and χ_b are defined in equations (3.38) and (3.57) for ferromagnets and uniaxial antiferromagnets, respectively. The inclusion of the scalar factor μ_{∞} is to account for the background permeability in a way that is analogous to the introduction of ε_{∞} in the scalar dielectric function (see section 4.3). In chapter 3 we had, for simplicity, taken $\mu_{\infty} = 1$.

From Maxwell's equations the magnetic-field vectors **H** and **M** satisfy

$$c^2\left[\nabla(\nabla \cdot \mathbf{H}) - \nabla^2\mathbf{H}\right] = -\varepsilon\frac{\partial^2}{\partial t^2}(\mathbf{H} + \mathbf{M}) \tag{5.26}$$

when retardation effects are included. Assuming now a dependence for the fluctuating field terms such as $\exp(i\mathbf{q} \cdot \mathbf{r} - i\omega t)$ for bulk waves in an unbounded magnetic material, where **q** is a 3D wave vector, the above equation becomes

$$q^2\mathbf{h} - \mathbf{q}(\mathbf{q} \cdot \mathbf{q}) = (\varepsilon\omega^2/c^2)(\mathbf{h} + \mathbf{m}). \tag{5.27}$$

Here **m** and **h** are defined as in equations (3.18) and (3.19) and ε is the dielectric constant of the material. If we now take coordinate axes such that the z-axis is along the static magnetization direction for a ferromagnet (or along the uniaxial direction for an antiferromagnet) and we choose $q_y = 0$ without loss of generality, then equation (5.27) can be rewritten in component form as

$$h_x + ih_y = \xi m^+ - \xi' m^-, \qquad h_x - ih_y = -\xi' m^+ + \xi m^-, \tag{5.28}$$

where

$$\xi = \frac{2(\varepsilon\omega^2/c^2) - q_x^2}{2\left[q^2 - \left(\varepsilon\omega^2/c^2\right)\right]}, \qquad \xi' = \frac{q_x^2}{2\left[q^2 - \left(\varepsilon\omega^2/c^2\right)\right]}. \tag{5.29}$$

When these results are combined with the susceptibility relations derived in subsection 3.3.1 they lead to the implicit dispersion relation for bulk magnetic polaritons (or magnon-polaritons) as

$$(1 - \xi\chi_a)^2 - \xi^2\chi_b^2 = (\xi')^2(\chi_a^2 - \chi_b^2). \tag{5.30}$$

This equation is applicable to both ferromagnets and antiferromagnets. In the ferromagnetic case, with χ_a and χ_b given by equation (3.38), the result becomes

$$\omega^2 = \omega_0^2 - 2\xi\omega_0\omega_M + \left[\xi^2 - (\xi')^2\right]\omega_M^2, \tag{5.31}$$

which is equivalent to an expression first given by Auld [29]. The two branches to the polariton spectrum, which depend on the angle θ between the propagation vector $\mathbf{q}$ and the z-axis, are illustrated by the numerical examples for $\theta = 0°$ and $90°$ in figure 5.11. The corresponding dispersion curves in the case of an antiferromagnet are discussed, for example, by Sarmento and Tilley [30].

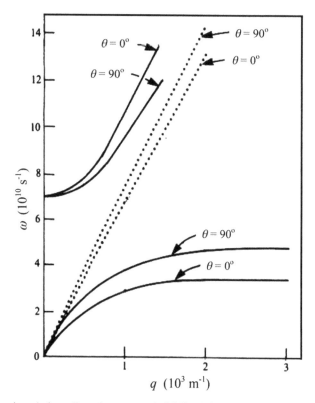

Figure 5.11. Magnetic polariton dispersion curves (solid lines) for a bulk ferromagnet for two different propagation angles $\theta = 0°$ and $90°$. The dashed lines are the light lines. The chosen parameters are $B_0 = \mu_0 M_0 = 0.175$ T. After Auld [29].

5.3.1 One- and two-interface geometries

With the results for bulk magnon-polaritons established we are now in a position to describe the behaviour of these polaritons when interfaces are involved. In this subsection we focus our attention on the geometries with just one or two interfaces. Specifically we will consider the film geometry depicted in figure 3.6, where the static magnetization along the z-axis is parallel to the interfaces (i.e., in the yz-plane). There are interfaces at $x = 0$ and $x = -L$, and in the semi-infinite limit ($L \to \infty$) only the $x = 0$ interface becomes of interest. The surface magnon-polaritons in this film geometry were first studied by Karsono and Tilley [31] and subsequently their analysis was extended to include both surface and guided magnon-polaritons by Marchand and Caillé [32].

Here we focus, for simplicity, on a ferromagnetic film in the Voigt geometry (the case of angle $\phi = 90°$ in figure 3.6, but the theory has also been carried out for antiferromagnets and for general values of ϕ. We now deal directly with the fluctuating fields **h** and **m** instead of the magnetostatic scalar potential as in subsection 3.3.1 where retardation was neglected. The analogous expressions to replace equation (3.43) in the three spatial regions are

$$
\mathbf{h} = \begin{cases}
\mathbf{a}_1 \exp(-\kappa_a x) \exp\!\big(iq_{\parallel} y - i\omega t\big) & \text{for } x > 0, \\
\big[\mathbf{a}_2 \exp(iq_x x) + \mathbf{a}_3 \exp(-iq_x x)\big] \exp\!\big(iq_{\parallel} y - i\omega t\big) & \text{for } 0 > x > -L, \\
\mathbf{a}_4 \exp(\kappa_a x) \exp\!\big(iq_{\parallel} y - i\omega t\big) & \text{for } x < -L,
\end{cases}
\tag{5.32}
$$

where the surrounding non-magnetic media are taken to be air or vacuum and

$$
\kappa_a = \sqrt{q_{\parallel}^2 - \big(\omega^2/c^2\big)}.
\tag{5.33}
$$

The quantity q_x may be real or imaginary in general and is found from

$$
q_x^2 = \varepsilon \mu_V(\omega) \big(\omega^2/c^2\big) - q_{\parallel}^2,
\tag{5.34}
$$

where $\mu_V(\omega)$ is an effective frequency-dependent permeability, sometimes known as the Voigt permeability, given in terms of equation (5.25) for the permeability tensor by

$$
\mu_V(\omega) = \mu_{xx} + \big(\mu_{xy}^2/\mu_{xx}\big).
\tag{5.35}
$$

The quantity q_x can either be imaginary (whereupon we denote $q_x = -i\kappa_b$ with $\kappa_b > 0$) or real, depending on the sign of the right-hand side of equation (5.34). These situations correspond, respectively, to surface modes (decaying in the film) and guided modes (wave-like in the film). The calculation for the surface modes is more straightforward and follows as a generalization of the DE magnetostatic calculation in subsection 3.3.1. It is necessary to apply the full electromagnetic boundary conditions for electric- and magnetic-field components at the $x = 0$ and $x = -L$

interfaces [31]. The implicit result for the surface magnon-polaritons of the film can eventually be written as (see [32])

$$\left(\kappa_a^2 \mu_V^2 + \kappa_b^2 + q_\parallel^2 \frac{\mu_{xy}^2}{\mu_{xx}^2} \right) \tanh(\kappa_b L) + 2\kappa_a \kappa_b \mu_V = 0. \tag{5.36}$$

The above result can be regarded as a generalization of equation (3.45) for the magnetostatic DE modes of a magnetic film. It is useful to look at some simplifications in the single-interface case, which is formally obtained if we take the limit of $L \to \infty$. The localization condition for existence of a surface mode is found to be (in the absence of damping)

$$\kappa_a \mu_V + \kappa_b = i q_\parallel \mu_{xy}/\mu_{xx}, \tag{5.37}$$

where we recall that μ_{xy} is imaginary so all the above terms are real. Equation (5.37) can be satisfied only in certain intervals of the frequency ω, which will be different depending on whether $q_y = +q_\parallel$ or $-q_\parallel$. This latter property is another manifestation of the non-reciprocal propagation that was mentioned in subsection 3.3.1 for the magnetostatic (DE) modes. Suppose we take the same sign conventions as before, which are consistent with having $\kappa_b > 0$. It is then found (see [28]) that when $|q_y| = q_\parallel \gg \omega/c$ there is a surface magnon-polariton localized near the $x = 0$ interface when $q_y > 0$ but no surface mode when $q_y < 0$. This is the regime where retardation effects are small and the surface-mode frequency is approximately given by

$$\omega_S \simeq \omega_0 + \frac{\mu_\infty \omega_M}{\mu_\infty + 1}, \tag{5.38}$$

which is just the result in equation (3.47) when $L \to \infty$ generalized to $\mu_\infty \neq 1$. On the other hand, when $q_\parallel \sim \omega/c$ the retardation effects are strong and it is found that surface magnon-polaritons may exist for both signs of q_y and they are still non-reciprocal.

The corresponding theory has also been developed for the interface modes in antiferromagnetic materials (see, e.g., [33]). Generally, the results for the magnon-polaritons are more complicated, but there is a simplification when the applied magnetic field is zero ($B_0 = 0$) because the off-diagonal susceptibility χ_b vanishes due to cancellations between the two sublattices, as can be seen from equations (3.57) and (3.58). As a consequence, the propagation of the modes becomes reciprocal in this special case and the dispersion relation of the surface magnon-polaritons is

$$\frac{c^2 q_\parallel^2}{\omega^2} = \frac{\mu_{xx}(\mu_{xx} - \varepsilon)}{\mu_{xx}^2 - 1}. \tag{5.39}$$

As mentioned in chapter 3 the FMR frequencies are in the microwave region, so typical wavelengths are several centimetres. It follows, therefore, that the study of the retarded magnon-polaritons in ferromagnets would require excessively large sample sizes. In contrast, the AFMR frequencies are often in the far-infrared, so experimental studies of antiferromagnetic magnon-polaritons are more feasible. The first experimental observation of magnon-polaritons was reported by Jensen *et al* [34] for FeF_2 using high-resolution reflectivity. Subsequently, ATR and improved reflectivity measurements on FeF_2 were reported [35]. A comparison of an ATR experiment and theory is shown in figure 5.12. The applied field of magnitude 0.3 T is reversed in direction in the two spectra and the different features clearly exhibit the non-reciprocity of the magnon-polaritons.

5.3.2 Multiple-interface geometries

We now extend some of the preceding calculations for magnon-polaritons in the one- and two-interface cases to a two-component superlattice using the same geometry as depicted in figure 3.17, except that in the present context the magnetic layers (constituent material *B*) can be considered as being either ferromagnetic or antiferromagnetic. It actually turns out to be straightforward to treat the general

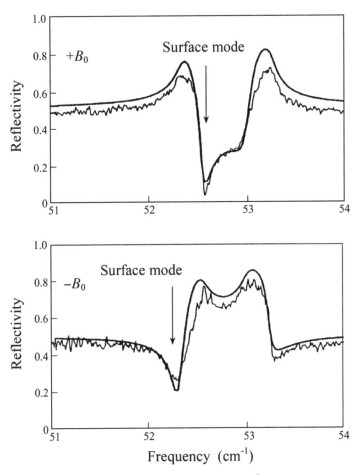

Figure 5.12. ATR spectra of FeF_2 at 1.7 K with resolution 0.02 cm^{-1} for two signs of the magnetic field $B_0 = \pm 0.3$ T. Experiment and theory are shown by the thinner and thicker lines. After Jensen *et al* [35].

case where both the A and B constituent materials can be magnetic, as described below.

The theory can be carried out using the transfer-matrix method as a direct extension of the magnetostatic case in subsection 3.6.2 to include the retardation effects (see, e.g., [36, 37]). Thus we deal with expressions for the spatial dependence of the electric- and magnetic-field components in the layers, instead of using the magnetostatic potential. Specifically, we consider s-polarized polaritons in the Voigt geometry, so the **E** field has only a single component E_z. In medium $J (=A, B)$ of the lth cell, defined as before, the field is given by an expression of the form

$$E_{Jz}^{(l)} = \left[a_J^{(l)} \exp\left(\kappa_J[x + (l-1)D]\right) + b_J^{(l)} \exp(-\kappa_J[x + (l-1)D]) \right] \exp\left(iq_\parallel y - i\omega t\right),$$

(5.40)

where

$$\kappa_J = \left[q_\parallel^2 - \varepsilon_J \mu_V^{(J)}(\omega)(\omega^2/c^2) \right]^{1/2}, \quad (J = A, B).$$

(5.41)

The analogous expressions for the components of the **H** can also be written down. The Voigt permeability $\mu_V^{(J)}$ of each material is defined as in equation (5.35).

After using the electromagnetic boundary conditions at each interface of cell l and following through with the calculation of the transfer matrix as described in subsection 1.5.1, the Rytov-type equation for the bulk magnon-polaritons of the superlattice becomes

$$\cos(QD) = \cosh(\kappa_A d_A)\cosh(\kappa_B d_B) + g \sinh(\kappa_A d_A)\sinh(\kappa_B d_B),$$

(5.42)

where

$$g = \frac{q_\parallel^2 \chi_b^{(A)} \chi_b^{(B)}}{\kappa_A \kappa_B} - \frac{q_\parallel \mu_V^{(A)}}{i\kappa_A}\Gamma_B - \frac{q_\parallel \mu_V^{(B)}}{i\kappa_B}\Gamma_A$$

(5.43)

and

$$\Gamma_J = \frac{\kappa_J^2\left(1 + \chi_a^{(J)}\right)^2 - q_\parallel^2\left(\chi_b^{(J)}\right)^2}{iq_\parallel \kappa_J \mu_V^{(J)}\left(1 + \chi_a^{(J)}\right)^2}, \quad (J = A, B).$$

(5.44)

This general result is algebraically relatively complicated, but there are certain special cases for which the expression for g simplifies. One such case of interest is when one of the materials (say, A) is non-magnetic, since we then have $\chi_a^{(A)} = \chi_b^{(A)} = 0$ and $\mu_V^{(A)} = 1$. A numerical example of the magnon-polariton dispersion relationships for this case is shown in figure 5.13, where the magnetic material (medium B) is the antiferromagnet MnF_2. Results are included here both for the bulk modes (shaded regions) and for the surface modes (dashed lines) of the superlattice [38]. In figure 5.13(a) where $B_0 = 0$ the surface mode exhibits reciprocal propagation, contrasting with figure 5.13(b) where $B_0 \neq 0$ and the surface-mode behaviour is seen to be non-reciprocal.

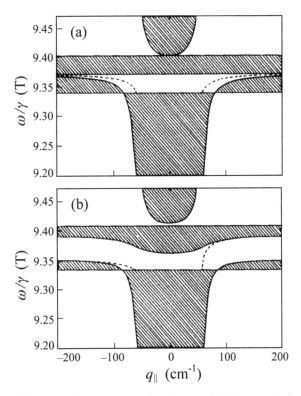

Figure 5.13. Calculated frequencies of magnon-polaritons for a semi-infinite superlattice of alternating layers of antiferromagnetic MnF_2 (medium B) and a non-magnetic material (medium A), showing bulk modes (shaded) and surface modes (dashed lines) when $d_A = d_B = 500$ μm. The two panels correspond to (a) $B_0 = 0$ and (b) $B_0 = 0.02$ T. After Barnás [38].

If the wave-vector propagation is different from the Voigt configuration, meaning angle $\phi \neq 90°$ in figure 3.6, the use of a generalized 4×4 transfer-matrix formalism [39] is necessary, since the x-dependence of the electric and magnetic fields is more complicated than in equation (5.40). Another approach, which simplifies the analysis, is to use an effective-medium approximation analogous to that described in section 5.1.3 for phonon-polaritons. This approach was explored by Raj and Tilley [40] (see also [8]).

5.4 Other types of polaritons

In the previous parts of this chapter we have covered the most important types of polaritons, i.e., those involving the coupling of light (photons) to phonons, plasmons and magnons. Other cases to be considered more briefly in this section are magnetoplasmon-polaritons and exciton-polaritons. Both involve the use of more complicated expressions for the dielectric functions. Looking ahead, piezoelectric materials are one of the topics in chapter 6, where further discussion of polaritons will be given in that context.

5.4.1 Magnetoplasmon-polaritons

As part of the discussion of the dielectric function of a charge plasma in section 4.3, we mentioned the modifications arising if a static applied field of magnitude B_0 is applied. In particular, the scalar ε becomes replaced by a gyrotropic tensor form defined by equations (4.28)–(4.30). This involves another characteristic frequency ω_c, the cyclotron frequency given by equation (4.31), as well as the plasma frequency ω_p.

The electromagnetic modes associated with this modified dielectric tensor are known as *magnetoplasmon-polaritons* and their properties at interfaces were discussed by Wallis *et al* [41] and Albuquerque *et al* [42], among others. Results depend on whether the applied field is oriented parallel or perpendicular to the interface planes, but the former case is usually adopted. The topic is thoroughly reviewed in [4], where a 2×2 transfer-matrix calculation is described for a two-component *ABABAB...* superlattice and for a four-component n–i–p–i superlattice.

Following [41] some bulk and surface magnetoplasmon-polariton dispersion relations are shown in figure 5.14 for a semi-infinite two-component superlattice. Here the axes are chosen such that $\mathbf{B}_0$ is in the y-direction and the z-axis is normal to the interfaces. The propagation of modes along the x-direction is studied, so essentially we have the Voigt configuration. The superlattice parameters correspond to $\omega_{pA} = 2\omega_{pB}$ and $d_A = 0.5d_B$, while the magnitude of the applied field is chosen so that the cyclotron frequency $\omega_c = 0.5\omega_{pA}$ in both materials. In the absence of an

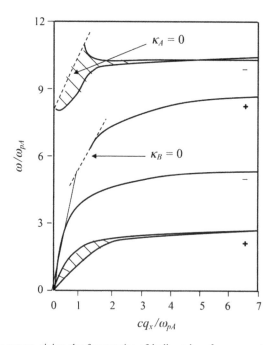

Figure 5.14. Dispersion curves giving the frequencies of bulk and surface magnetoplasmon-polaritons in a semi-infinite two-component superlattice as a function of in-plane wave vector q_x. Bulk-mode regions are shown shaded and the four surface modes are indicated by solid lines with plus or minus labels. See the text for explanation and for parameter values. After Wallis *et al* [41].

applied magnetic field it can be shown (e.g., from the formalism of section 5.2) that there would be two surface plasmon-polariton branches, doubly degenerate and corresponding to propagation along the $\pm x$-directions. The effect of $\mathbf{B}_0$ is to remove this degeneracy, leading to the four surface magnetoplasmon-polariton branches labelled with the plus and minus signs in the figure. One of these branches occurs just below the lower bulk band at larger q_x and merges into it when $cq_x/\omega_{pA} \simeq 3$. The other three surface modes lie between the two bulk bands and one of these has the unusual property that it ceases to exist (becomes unlocalized) for an intermediate q_x range when it meets the $\kappa_B = 0$ line (shown dashed).

5.4.2 Exciton-polaritons

The concept of excitons was introduced in section 4.3, where it was explained that spatial dispersion effects are large (due to the exciton kinetic energy) and so they have to be taken into account leading to the dielectric function $\varepsilon(q, \omega)$ given in equation (4.39). When this dielectric function is substituted into equation (1.20) we obtain the bulk exciton-polariton dispersion relation. It consists of two branches, by analogy with the phonon-polariton case in figure 5.1, but has novel features due to the q-dependence of $\varepsilon(q, \omega)$.

In particular, if the bulk frequencies are plotted as a function of q the exciton-polariton analogue of the lower branch in the phonon-polariton case will not simply flatten out at large q as in figure 5.1, but instead it will continue to curve upwards. The outcome is that, for any chosen frequency above the exciton frequency ω_e defined in section 4.3, there will be *two* possible values of q (see, e.g., [8]). This property has some important consequences when a boundary with another medium is introduced. To illustrate this, we suppose s-polarized light with $\omega > \omega_e$ is incident from a vacuum onto the planar surface of an excitonic material such as CdS or GaN. There will be a reflected wave and two transmitted transverse waves at different wave vectors. Maxwell's equations, however, provide us with just two boundary conditions, so an ABC is required. This need for an ABC is somewhat similar to the situation for SWs in the macroscopic dipole-exchange theory (see subsection 3.3.2). For exciton-polaritons a phenomenological ABC is usually adopted with the form [43]

$$\frac{\partial \mathbf{P}}{\partial z} + \xi \mathbf{P} = 0, \tag{5.45}$$

where $\mathbf{P}$ is the excitonic polarization, z is the coordinate axis perpendicular to the interface and ξ is a constant. This expression has some formal similarities with the ABC quoted in equation (3.67) for the magnetic case. Another simple form was $\mathbf{P} = 0$ for the excitonic polarization at the surface ($z = 0$) as proposed in [44]. Later work has helped to put the above forms of ABC on a firmer theoretical basis (see discussion in [4]).

A calculation for the surface and guided exciton-polaritons in a GaN thin film [45] is shown in figure 5.15. The $\mathbf{P} = 0$ form for the ABC was used at both interfaces and damping was ignored in the dielectric function. Among the techniques for observing these modes are photoreflectance and photoluminescence, and the above theory was compared with experiments by Gil *et al* [46] giving reasonable agreement.

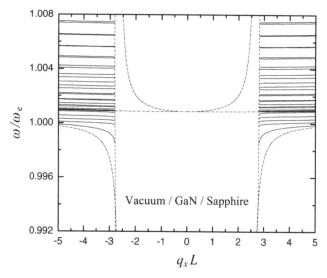

Figure 5.15. Dispersion curves for p-polarized exciton-polaritons in a 50 nm GaN film bounded by a vacuum and sapphire. The full lines represent the calculated surface and guided modes. The near-vertical lines are the light lines and the chain-dashed lines are two transverse and one longitudinal bulk mode. After Vasconcelos *et al* [45].

The full calculations for exciton-polaritons in superlattices are extremely complicated, but in practice an effective-medium approach similar to that described earlier in this chapter can be employed (see, e.g., [17, 47]).

References

[1] Burstein F and de Martini F (ed) 1974 *Polaritons* (New York: Pergamon)
[2] Agranovich V M and Mills D L (ed) 1982 *Surface Polaritons* (Amsterdam: North-Holland)
[3] Boardman A D (ed) 1982 *Electromagnetic Surface Modes* (Chichester: Wiley)
[4] Albuquerque E A and Cottam M G 2004 *Polaritons in Periodic and Quasiperiodic Structures* (Amsterdam: Elsevier)
[5] Otto A 1976 *Optical Properties of Solids: New Developments* ed B O Seraphim (Amsterdam: North-Holland) p 678
[6] Kawata S (ed) 2001 *Near-Field Optics and Surface Plasmon Polaritons* (Berlin: Springer)
[7] Kittel C 2004 *Introduction to Solid State Physics* 8th edn (New York: Wiley)
[8] Cottam M G and Tilley D R 2005 *Introduction to Surface and Superlattice Excitations* 2nd edn (IOP: Bristol)
[9] Oliveira F A, Cottam M G and Tilley D R 1981 *Phys. Status Solidi* B **107** 737
[10] Fukui M, So V C Y and Normandin R 1979 *Phys. Status Solidi* B **91** K61
[11] Sarid D 1981 *Phys. Rev. Lett.* **47** 1927
[12] Wendler L and Haupt R 1986 *Phys. Status Solidi* B **137** 269
[13] Marschall N and Fischer B 1972 *Phys. Rev. Lett.* **28** 811
[14] Sambles J R, Bradberry G W and Yang F 1991 *Contemp. Phys.* **32** 173
[15] Evans D J, Ushioda S and McMullen J D 1973 *Phys. Rev. Lett.* **31** 372
[16] Raj N and Tilley D R 1985 *Solid State Commun.* **55** 373

[17] Agranovich V M and Kravstov V E 1985 *Solid State Commun.* **55** 85
[18] Liu W M, Eliasson G and Quinn J J 1985 *Solid State Commun.* **55** 533
[19] Raj N, Camley R E and Tilley D R 1987 *J. Phys. C: Solid State Phys.* **20** 5203
[20] Bergman D J and Stroud D 1992 *Solid State Physics* vol 46 ed H Ehrenreich and D Turnbull (New York: Academic) p 147
[21] Maslin K A, Parker T J, Raj N, Tilley D R, Dobson P J, Hilton D and Foxon C T B 1986 *Solid State Commun.* **60** 461
[22] Dumelow T, Parker T J, Smith S R P and Tilley D R 1993 *Surf. Sci. Rep.* **17** 151
[23] Constantinou N C and Cottam M G 1986 *J. Phys. C: Solid State Phys.* **19** 739
[24] Qin G, Giuliani G F and Quinn J J 1983 *Phys. Rev.* B **28** 6144
[25] Farias G A, Auto M M and Albuquerque E L 1988 *Phys. Rev.* B **38** 12540
[26] Vollath D 2008 *Nanomaterials* (Weinheim: Wiley)
[27] Johnson B L and Camley R E 1988 *Phys. Rev.* B **38** 3311
[28] Hartstein A, Burtstein E, Maradudin A A, Brewster R and Wallis R F 1973 *J. Phys. C: Solid State Phys.* **6** 1266
[29] Auld B A 1960 *J. Appl. Phys.* **31** 1642
[30] Sarmento E F and Tilley D R 1975 *J. Phys. C: Solid State Phys.* **10** 795
[31] Karsono A D and Tilley D R 1978 *J. Phys. C: Solid State Phys.* **11** 3487
[32] Marchand M and Caillé A 1980 *Solid State Commun.* **34** 827
[33] Abraha K and Tilley D R 1996 *Surf. Sci. Rep.* **24** 125
[34] Jensen M R F, Parker T J, Abraha K and Tilley D R 1995 *Phys. Rev. Lett.* **75** 3756
[35] Jensen M R F, Feiven S A, Parker T J and Camley R E 1997 *J. Phys.: Condens. Matter.* **9** 7233
[36] Barnás J 1988 *J. Phys. C: Solid State Phys.* **21** 1021
[37] Raj N and Tilley D R 1989 *The Dielectric Function of Condensed Systems* ed I V Keldysh, D A Kirzhnitz and A A Maradudin (Amsterdam: Elsevier)
[38] Barnás J 1990 *J. Phys.: Condens. Matter* **2** 7173
[39] Barnás J 1991 *Phys. Status Solidi* B **165** 529
[40] Raj N and Tilley D R 1987 *Phys. Rev.* B **36** 7003
[41] Wallis R F, Szenics R, Quinn J J and Giuliani G F 1987 *Phys. Rev.* B **36** 1218
[42] Albuquerque E L, Fulco P, Farias G A, Auto M M and Tilley D R 1991 *Phys. Rev.* B **43** 2032
[43] Agranovich V M and Ginzburg V L 1962 *Sov. Phys.—Usp* **5** 323 and 675
[44] Pekar S I 1958 *Sov. Phys.—JETP* **7** 813
[45] Vasconcelos M S, Albuquerque E L, Farias G A and Freire V N 2002 *Solid State Commun.* **124** 109
[46] Gil B, Clur S and Briot O 1997 *Solid State Commun.* **104** 267
[47] Albuquerque E L and Fulco P 1994 *Z. Phys.* B **100** 289

IOP Publishing

Dynamical Properties in Nanostructured and Low-Dimensional Materials

Michael G Cottam

Chapter 6

Mixed excitations

In the last several chapters we have covered the main types of excitations in solids, along with their properties at surfaces and interfaces. Nevertheless there are some important areas not yet touched upon or only mentioned briefly and so it is useful to provide an account here in terms of a general survey with illustrative examples.

Some of the examples are for 'mixed' types of excitation that involve coupled waves of two different kinds. These include *magnetoelastic waves*, where there is a coupling between magnetic and elastic vibrational modes, and *piezoelectric waves*, where the electric-field properties are coupled to crystal deformations. There will also be discussion of *ferroelectrics*, i.e., materials that may exhibit a spontaneous electric-dipole moment at low temperatures, just as ferromagnets have a spontaneous net magnetization. This topic leads into a class of materials known as *multiferroics*, where ferromagnetic and ferroelectric properties may be exhibited simultaneously (and may sometimes be coupled to one another).

6.1 Magnetoelastic waves

Magnetoelastic waves arise from a coupling between the magnetic and vibrational (phonon) modes when the latter are treated in terms of an elastic continuum. The form taken by the magnetoelastic interaction was first considered by Turov and Irkhin [1] and Kittel [2] using phenomenological arguments, but a detailed derivation for the interaction Hamiltonian came shortly afterwards due to Schlömann [3] who found a combination of four terms:

$$H_{m-e} = \frac{B_1}{M_0^2} \sum_{\mu} M_\mu^2 \bar{u}_{\mu\mu} + \frac{B_2}{M_0^2} \sum_{\mu,\nu(\neq\mu)} M_\mu M_\nu \bar{u}_{\mu\nu}$$

$$+ \frac{A_1}{M_0^2} \sum_{\mu,\nu,\eta(\neq\nu)} \frac{\partial M_\mu}{\partial r_\mu} \frac{\partial M_\mu}{\partial r_\eta} \bar{u}_{\nu\eta} + \frac{A_2}{M_0^2} \sum_{\mu,\nu} \left(\frac{\partial M_\mu}{\partial r_\mu}\right)^2 \bar{u}_{\nu\nu}. \tag{6.1}$$

doi:10.1088/978-0-7503-1054-3ch6

The notation is consistent with that in chapters 2 and 3, with $\bar{u}_{\mu\nu}$ and M_μ being Cartesian components of the strain tensor (see section 2.2) and total magnetization (see subsection 3.2), respectively. The summations are over x-, y- and z-coordinates, and the partial derivatives are with respect to spatial components. The first two terms, proportional to B_1 and B_2, represent the dependence of the magnetic dipole–dipole and spin–orbit interactions, respectively, on the elastic strain. The coefficients are expected to be comparable in magnitude; in fact, it can be deduced that $B_1 = B_2$ for a ferromagnet with assumed isotropic elastic properties. In practice these coefficients may be different, e.g., for the ferrimagnet YIG measurements have shown that $B_1 \simeq 0.5B_2$. The last two terms in equation (6.1), proportional to A_1 and A_2, are due to modulations of the Heisenberg exchange term by the strain. Their main effect is to describe magnon–phonon processes in which two magnons interact with a phonon and they are more appropriately considered along with other NL effects (see chapter 7). All of the terms in equation (6.1) can also provide contributions to the effective magnetic anisotropy at the bulk and surface of the material (see, e.g., [4]).

At sufficiently long excitation wavelengths, as discussed in chapter 3, the exchange effects become unimportant. It can then be shown that the main effect of the magnetoelastic coupling in the dynamical properties comes from the B_2-term in equation (6.1). The dominant contributions in the summation are those involving products such as $M_z M_x \bar{u}_{zx}$, which yields $M_0 m_x \bar{u}_{zx}$ after a linearization approximation. Hence a simplified form for the magnetoelastic interaction when the anisotropy energy is neglected is

$$H_{m-e} = \frac{B_2}{M_0}\Big[m_x(\bar{u}_{xz} + \bar{u}_{zx}) + m_y(\bar{u}_{yz} + \bar{u}_{zy})\Big]. \tag{6.2}$$

This is just the same as the expression introduced phenomenologically by Kittel [2]. Apart from the references already given, some other introductory accounts of magnetoelastic interactions and waves can be found in [5–8].

6.1.1 Bulk waves

We can derive the dispersion relation(s) for the magnetoelastic waves by writing down the appropriate equations of motion for the four coupled quantities m_x, m_y, u_x and u_y. The total Hamiltonian consists of H_{m-e} together with the equations for non-interacting elastic waves (phonons) and magnons. The calculation will be performed first for a bulk medium where we seek solutions with a plane-wave form such as $\exp(\mathrm{i}\mathbf{q} \cdot \mathbf{r} - \mathrm{i}\omega t)$ as usual. Using equation (6.2) and taking the case of propagation wave vector $\mathbf{q}$ along the z-direction of magnetization in a cubic ferromagnet, we find

$$\mathrm{i}\omega m_x + \omega_{\mathrm{mag}}(q)m_y = \mathrm{i}qB_2 u_y, \qquad \mathrm{i}\omega m_y - \omega_{\mathrm{mag}}(q)m_x = -\mathrm{i}qB_2 u_x,$$

$$\omega^2 \rho u_x - q^2 \lambda_{44} u_x = \mathrm{i}q\frac{B_2}{M_0}m_x, \qquad \omega^2 \rho u_y - q^2 \lambda_{44} u_y = \mathrm{i}q\frac{B_2}{M_0}m_y. \tag{6.3}$$

Here the non-interacting magnon and phonon dispersion relations are $\omega_{\mathrm{mag}}(q) = \omega_0 + Dq^2$ and $\omega_{\mathrm{ph}} = v_{\mathrm{T}}q = (\lambda_{44}/\rho)^{1/2}q$, respectively. The notation is consistent with chapters 2 and 3, and we note in particular that v corresponds to the velocity of TA phonons with λ_{44} denoting one of the elastic constants of a cubic material [9]. Instead of the transverse m_x and m_y variables we could alternatively have employed the circular combinations $m^+ = m_x + im_y$ and $m^- = m_x - im_y$ by analogy with chapter 3. The mode coupling occurs when $B_2 \neq 0$ leading to the extra terms on the right-hand side of each equation in (6.3). We could add another equation, the wave equation for u_z, to those in (6.3), but it is uninteresting because it does not couple to the magnons in the linear approximation.

From the condition for the above four coupled equations to have consistent and non-trivial solutions we deduce that ω satisfies

$$\left[\omega \mp \omega_{\mathrm{mag}}(q)\right]\left[\omega^2 - \omega_{\mathrm{ph}}^2(q)\right] \mp bq^2 = 0, \tag{6.4}$$

where $b = B_2^2/\rho M_0$ is an effective coupling coefficient. The upper and lower signs in equation (6.5) refer to the waves with a right-hand and left-hand sense of circular polarization, respectively. The form of the dispersion relations can be explained using figure 6.1. In the absence of magnetoelastic coupling ($b = 0$) it is clear that the unperturbed modes at positive frequency are $\omega = \omega_{\mathrm{mag}}(q)$ and $\omega = \omega_{\mathrm{ph}}(q)$; these are the dashed lines in the figure. For small enough applied magnetic field these lines have two points of intersection as indicated by the vertical blue arrows. The solid lines in figure 6.1 show the magnetoelastic modes obtained when $b \neq 0$. For the right-handed polarization there are two solutions (labelled as 'RH1' and 'RH2')

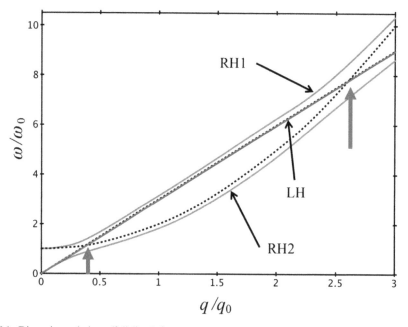

Figure 6.1. Dispersion relations (full lines) for transverse-polarized bulk magnetoelastic modes in a cubic ferromagnet for the case of propagation along the magnetization direction. Here $q_0 = (\omega_0/D)^{1/2}$ is a characteristic wave vector and we have assumed $v_T q_0 = 3\omega_0$ and $b = D\omega_0^2$ for illustration. The right-hand magnetoelastic modes are labelled 'RH1' and 'RH2', while the left-hand mode is 'LH'. The non-interacting magnon and phonon lines are shown dashed and they intersect at points indicated by the vertical blue arrows.

which exhibit a typical 'mode repulsion' behaviour near the intersection points. For the left-handed polarization there is just one solution (labelled as 'LH') which lies very close to the transverse phonon line.

The bulk magnetoelastic modes can similarly be found for other propagation directions, but they are slightly more complicated because the dipole–dipole terms contribute. For example, suppose we now take the case of propagation wave vector **q** perpendicular to the magnetization direction and along one of the cubic symmetry axes. After deriving a set of coupled equations analogous to those in equation (6.3) we find that the magnetoelastic spectrum consists of just one branch with its implicit dispersion relation given by

$$\left[\omega^2 - \omega_{mag}^2(q)\right]\left[\omega^2 - \omega_{ph}^2(q)\right] - b\omega_M q^2 = 0, \tag{6.5}$$

where b is the same as before. Here ω_M is related to the magnetization as in equation (3.39) and the unperturbed magnon dispersion is now given by the modified expression

$$\omega_{mag}(q) = \left[\left(\omega_0 + Dq^2\right)\left(\omega_0 + \omega_M + Dq^2\right)\right]^{1/2}, \tag{6.6}$$

which is equivalent to the bulk dipole-exchange result obtained in equation (3.66). Discussion of the solution of equation (6.5), along with properties of magnetoelastic waves propagating in lower-symmetry directions, can be found in [8].

6.1.2 Surface and superlattice waves

Next we consider the magnetoelastic surface wave in the single-interface case of a semi-infinite ferromagnetic material. In principle, there are now several possible schemes whereby the unperturbed magnetic and elastic modes could be coupled when $B_2 \neq 0$. The relevant modes are those discussed in chapters 2 and 3, which include the surface and bulk magnetostatic or dipole-exchange modes on the magnetic side together with the bulk and surface (Rayleigh) elastic waves on the other. Typically, the 'mixed' magnetoelastic surface modes will exhibit non-reciprocal propagation for geometries where the magnetization is parallel to the interface, since this is a characteristic of the unperturbed surface magnetic modes (see chapter 3).

As an example for the single-interface case we will briefly discuss a calculation by Scott and Mills [10]. They studied a ferromagnetic material in the geometry of figure 3.6, but with $L \to \infty$ for the semi-infinite case. The magnetization was parallel to the interface plane and surface acoustic-type waves, as well as magnetostatic waves, could propagate in the material. The other half-space with $x > 0$ was considered to be a vacuum. They assumed a form for the magnetoelastic coupling H_{m-e} equivalent to equation (6.2). Essentially, by combining the same theoretical techniques and the boundary conditions as used earlier in subsections 2.2.1 and 3.3.1, they carried out calculations for two generalizations of the surface acoustic waves in the presence of the magnetoelastic coupling. One calculation was for the Rayleigh surface wave modified by the H_{m-e} term and the other was for a shear-like

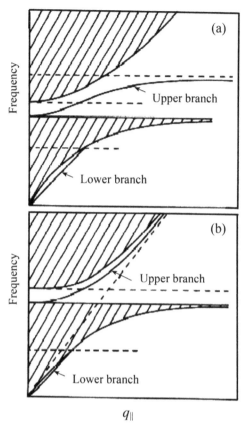

Figure 6.2. Frequency of shear-polarized magnetoelastic surface waves (solid lines) as a function of the in-plane wave vector $q_{\parallel}$. The Voigt geometry is assumed taking (a) $\phi = 90°$ and (b) $\phi = -90°$. The bulk continuum is shown shaded. After Scott and Mills [10].

magnetoelastic wave that exists only when $H_{m-e} \neq 0$. Some results for the dispersion relations are shown in figure 6.2, where we show two cases for the angle ϕ (see figure 3.6) corresponding to reversing the propagation wave vector $\mathbf{q}_{\parallel}$ in the Voigt geometry. The non-reciprocity properties of the surface modes (with two branches) are evident. In each case for ϕ the lower surface branch is analogous to the Rayleigh surface wave, while the upper surface branch is a distinctively new feature. Scott and Mills concluded that the magnetoelastic coupling could provide a sensitive experimental technique for probing (e.g., via the Rayleigh waves) the magnetic modes.

Experimental studies of the surface and bulk magnetoelastic waves in thin films of the ferrimagnet YIG were, for example, made by Temiryazev *et al* [11]. As well as requiring a generalization of the preceding theory to thickness L being finite, their work also brought in new effects due to the strong exchange interactions in YIG. Other calculations have been reported for magnetoelastic waves in ferromagnetic films (see, e.g., [12]) and at planar interfaces between two different ferromagnets [13].

There have also been extensions, mostly on the theoretical side, to study magnetoelastic waves in superlattices. Initially the emphasis was on using the effective-medium approximation, which was described for other excitations in sections 5.1 and 5.3. As mentioned there, it is a useful approximation if the superlattice periodic length D is sufficiently small compared with any of the characteristic wavelengths for the excitation. Detailed calculations applicable under these conditions have been made by Tarasenko and co-workers (see, e.g., [14] and references therein). For two-component superlattices with an alternating $ABABAB...$ structure and larger D values, Belykh *et al* [15] have employed a transfer-matrix approach analogous to that described in section 1.5 and subsequently. Their formalism included the superlattice surface modes in finite and semi-infinite structures as well as the superlattice bulk modes of an infinitely extended structure.

So far in this section we have focused on magnetoelastic waves in the context of ferromagnetic materials, but there has also been interest regarding antiferromagnetic and paramagnetic materials. The extension to antiferromagnets is rather straightforward since the main difference in the theory is that the magnetic susceptibility terms χ_a and χ_b (see subsection 3.3.1), which are used in obtaining the coupled equations in (6.3), have to be replaced by the corresponding susceptibility expressions for antiferromagnets. One noteworthy consequence of this, however, is that in the special case of zero applied magnetic field we have the simplification that $\chi_b = 0$ and so the antiferromagnetic magnetoelastic surface modes have reciprocal propagation properties.

The case of magnetoelastic modes in paramagnetic materials (such as rare-earth compounds) is rather simpler in principle, since there is no static magnetization or sublattice magnetization. Hence the unperturbed magnon branch is simply the dispersionless $\omega = \omega_0$. The magnetoelastic interactions arise from the coupling between strain and rotational terms in the deformation tensor of the unfilled $4f$ electron shell. Typically this results in the elastic properties of the materials showing a strong dependence on temperature and applied magnetic field. These effects can be probed very effectively through the properties of the Rayleigh wave using a single-interface geometry. Experimental observations of the behaviour of the Rayleigh wave in paramagnetic $CeAl_2$ and $SmSb$ were reported by Lingner and Lüthi [16], giving good agreement with the magnetoelastic mode theory.

6.2 Piezoelectric waves

Piezoelectrics are a class of materials with the property that when they are deformed they give rise to an electric field. Conversely, when an electric field is applied to a piezoelectric material a deformation takes place. To describe a wave propagating in such a material we require a combination of the elastic equations of motion in the material together with Maxwell's equations of electromagnetism. Both of these have to be augmented by terms describing the piezoelectric coupling, which has a tensorial character that depends on the symmetry of the material.

Quartz is a well-known example of a piezoelectric material; others include materials with non-centrosymmetric lattices such as lead titanate ($PbTiO_3$), barium titanate ($BaTiO_3$) and lead zirconate titanate ($PbZr_xTi_{1-x}O_3$ with $0 < x < 1$, usually called PZT). The electric field produced when a surface acoustic wave propagates on a piezoelectric material may be employed to couple to an external circuit or to other surface waves that have an electromagnetic component or to the charge carriers in a semiconductor. Hence, together with the inverse effect, there are many possibilities for practical applications of piezoelectricity in the general area of acousto-electronics. Some references for examples and further details of applications are [17, 18].

The basic results for the development of piezoelectric waves are as follows (see, e.g., [19]). First there is the generalization of Hooke's law in section 2.2 to give an expression for the stress σ_{ij} as

$$\sigma_{ij} = \sum_{kl} \lambda_{ijkl} \bar{u}_{kl} - \sum_{k} e_{kij} E_k, \tag{6.7}$$

where λ_{ijkl} and $\bar{u}_{kl}$ are elements of the elasticity tensor and strain, respectively, as before. The last term in equation (6.7) describes the coupling to the electric field where e_{kin} is a component of the piezoelectric tensor. The summations are over Cartesian components x, y and z. The inverse effect is that the electric displacement **D** is now given by

$$D_i = \varepsilon_0 \sum_{j} \varepsilon_{ij} E_j + \sum_{jk} e_{ijk} \bar{u}_{kl}, \tag{6.8}$$

which again involves the piezoelectric tensor. In addition to the above relationships, when solving for surface- and interface-wave problems, we require the use of appropriate electromagnetic and elastic boundary conditions. Some review accounts of surface piezoelectric waves are to be found in [20–22].

6.2.1 One-interface geometry

The first calculations made for the surface waves at the surface of a semi-infinite piezoelectric medium were essentially modifications of surface acoustic waves (e.g., see [23] and references therein). An interesting development leading to a new type of surface mode was reported independently by Bleustein [24] and Gulyaev [25]. This so-called Bleustein–Gulyaev (BG) wave has the unusual property that there is no direct analogue in the purely elastic system. Other surface-wave calculations were later performed by Albuquerque [26, 27] in a more general context to include other crystal symmetries and other orientations of the surface plane relative to the crystal axes. A surface response-function (or Green-function) method, which enables the relative intensities of the various piezoelectric modes to be deduced, was outlined in [21].

A simplification incorporated into the majority of calculations for piezoelectric waves involves the use of the *static-field approximation*, which is based on the fact that the electromagnetic-wave component of the mixed wave has a velocity that is several orders of magnitude larger than that of the acoustic-wave (elastic-wave) component. Hence the approximation is made that the electric field can be expressed in terms of an electrostatic scalar potential V as $\mathbf{E} = -\nabla V$. Then the coupled equations of motion consistent with the constitutive relations (6.7) and (6.8) become (see, e.g., [28])

$$\rho \frac{\partial^2 u_j}{\partial t^2} - \sum_{ikl} \lambda_{ijkl} \frac{\partial^2 u_k}{\partial r_i \partial r_l} = \sum_{ik} e_{kij} \frac{\partial^2 V}{\partial r_i \partial r_k},$$

$$\sum_{ik} \varepsilon_{ik} \frac{\partial^2 V}{\partial r_i \partial r_k} = \sum_{ikl} e_{ikl} \frac{\partial^2 u_k}{\partial r_i \partial r_l}. \tag{6.9}$$

As before, the summations and partial derivatives are with respect to Cartesian coordinates. Both equations in (6.9) have been arranged so that the piezoelectric coupling terms lead to a nonzero right-hand side. Depending on the orientation of the surface relative to the crystal axes, the boundary conditions might typically consist of requiring some component(s) of the stress tensor to vanish at the surface and for the potential V to satisfy a surface condition, plus there would be a condition for mode localization (decay as the distance into the material becomes large).

As an example, we start with the situation that gives rise to the BG wave mentioned above. In this case the surface is taken to be the plane $y = 0$. The piezoelectric material, chosen to have symmetry in the class C_{6v} (or $6mm$) in standard group-theoretical notations [29] fills the half-space $y \geqslant 0$. The six-fold axis is taken to coincide with the z-axis and the external medium (where $y < 0$) is air or a vacuum. The relevant equations, which are a subset of those in (6.9), involve the component u_z of the displacement and its coupling to V. They lead to

$$\rho \frac{\partial^2 u_z}{\partial t^2} - \lambda_{44} \nabla_2^2 u_z = e_{15} \nabla_2^2 V, \qquad \varepsilon_{xx} \nabla_2^2 V = e_{15} \nabla_2^2 u_z, \tag{6.10}$$

where ∇_2^2 is shorthand for the 2D Laplacian operator $\partial^2/\partial x^2 + \partial^2/\partial y^2$ while λ_{44} and e_{15} are elements of the elastic and piezoelectric tensors in a conventional contracted notation (see, e.g., [9, 28]).

Bleustein [24] considered the surface of the piezoelectric material at $y = 0$ to be coated with an infinitesimally thin, perfectly conducting electrode which was grounded. The problem is therefore to find surface-wave solutions of equation (6.10) subject to the boundary conditions that $\sigma_{yx} = \sigma_{yy} = \sigma_{yz} = 0$ at $y = 0$ for stress-free surface conditions, $V = 0$ at the grounded surface $y = 0$, and both $u_z \to 0$ and $V \to 0$ as $y \to \infty$ for localization.

The method of solution is facilitated if we define $W = V - (e_{15}/\varepsilon_{xx})u_z$, allowing equation (6.10) to be recast as

$$\rho \frac{\partial^2 u_z}{\partial t^2} - \Lambda_{44} \nabla_2^2 u_z = 0, \qquad \nabla_2^2 W = 0, \tag{6.11}$$

where $\Lambda_{44} = \lambda_{44} + (e_{15}^2/\varepsilon_{xx})$ is known as the 'piezoelectrically stiffened' elastic constant. It now follows that the surface-wave solutions for u_z and W must take the form

$$u_z = A_1 \exp(-\alpha y)\exp(iq_x x - i\omega t), \qquad W = A_2 \exp(-|q_x|y)\exp(iq_x x - i\omega t), \tag{6.12}$$

where A_1 and A_2 are constants and the positive quantity α satisfies

$$\rho\omega^2 = \Lambda_{44}\left(q_x^2 - \alpha^2\right). \tag{6.13}$$

Application of the boundary conditions leads to the extra relationship that $\alpha = |q_x|e_{15}^2/\varepsilon_{xx}\Lambda_{44}$, from which it may be deduced that the phase velocity $v_{\rm BG}$ of the BG surface wave satisfies

$$v_{\rm BG}^2 = \frac{\Lambda_{44}}{\rho}\left(1 - \frac{e_{15}^2}{\varepsilon_{xx}\Lambda_{44}}\right). \tag{6.14}$$

If we examine the limit of vanishing piezoelectric coupling ($e_{15} \to 0$) we see that $\alpha \to 0$ and so there is no longer a localized mode in this case.

Other types of surface modes were analysed by Albuquerque [26, 27] and reviewed in [22]. He examined the dispersion relations and scattering properties of surface waves, again for a semi-infinite piezoelectric material, assuming either cubic or hexagonal symmetries. Attention was given to different polarizations, leading to mixed waves with other components of the displacement being involved by comparison with the BG case.

6.2.2 Superlattices

Superlattices where at least one of the components is a piezoelectric material have also received attention. Generally the static-field approximation has been employed as in the previous subsection and it is well justified since the focus has mainly been on modifications to the acoustic (elastic) modes. In contrast, the case of the *retarded* piezoelectric polaritons has been less studied.

Some of the earlier superlattice calculations, such as those reported by Akcakaya and co-workers [30, 31], made use of an effective-medium approximation analogous to that described for superlattice polaritons in chapter 5. These authors concentrated on semi-infinite, two-component superlattices where both component materials were piezoelectric. They were concerned with modifications to the acoustic-type modes. The validity of their approach in using effective elastic and piezoelectric constants for the superlattice required the acoustic wavelength to be large compared with the periodic length of the superlattice.

In contrast, Albuquerque and Cottam [32] later employed a transfer-matrix approach (see below) which is a generalization of that described in section 1.5. They were concerned with superlattices that could be either periodic or quasi-periodic (following a Fibonacci sequence in this case) and were composed of two materials A and B, where only the latter is taken to be piezoelectric. They also made use of the static-field approximation, so that the basic relationships involve the scalar potential V and the components of the displacement $\mathbf{u}$ as in equation (6.9).

We adopt the same geometry and choice of coordinate axes as for the superlattice optical waves and acoustic waves in chapters 1 and 2, respectively, i.e., the z-axis is perpendicular to the interfaces and the wave propagation is along the x-direction. For the cubic and hexagonal classes of piezoelectric materials being studied in [32] it turns out that the coupled equations in (6.9) lead to the pairs (u_x, u_z) and (u_y, V) satisfying separate equations, with the coupling between the latter pair coming specifically from the piezoelectric terms. We seek wave-like solutions for u_y and V in the piezoelectric layers that can be cast in the form $\exp(iq_z z)\exp(iq_x x - i\omega t)$, where q_z may be complex in general. It can then be shown that there are actually two solutions for q_z, so a superposition has to be formed. It is eventually found [32], on dropping the common $\exp(iq_x x - i\omega t)$ factor, that we have

$$u_y = B_1 \exp(iqz) + B_2 \exp(-iqz),$$
$$V = (e_{x5}/\varepsilon_{xx})[B_1 \exp(iqz) + B_2 \exp(-iqz)] + B_3 \exp(-q_x z) + B_4 \exp(q_x z), \tag{6.15}$$

taking the case of hexagonal symmetry (e.g., ZnO). The B-coefficients can be determined later through application of the boundary conditions and the values of the effective wave vector q_z have been deduced to be iq_x and q where

$$q = \left[\left(q_{Tz}^2 - q_x^2 p \right) / (1 + p) \right]^{1/2} \tag{6.16}$$

and $p = e_{x5}^2 / \lambda_{44}\varepsilon_{xx}$ represents a dimensionless piezoelectric coupling coefficient. The z-component of the TA wave vector is

$$q_{Tz} = \left[(\omega/v_T)^2 - q_x^2 \right]^{1/2}, \qquad v_T = (\lambda_{44}/\rho)^{1/2}. \tag{6.17}$$

When the analogous expressions to equations (6.15) and (6.16) are written down for piezoelectric materials with cubic symmetry (e.g., GaAs) it is again found that there are combinations of four terms in u_y and V, although the results are slightly more complicated [32]. The consequence for the transfer-matrix method in both symmetry cases, however, is that 4×4 matrices must be employed instead of 2×2 matrices as previously. The appropriate 4×4 transfer matrices were calculated in [32] and employed to discuss the piezoelectric modes (see also [22] for a review). A numerical example is given in figure 6.3 for a periodic semi-infinite superlattice, showing bulk modes (shaded) and surface modes (dashed lines). The piezoelectric layers are assumed to be the hexagonal material ZnO and the non-piezoelectric layers are fused silica. The boundaries of the bulk-mode regions correspond to $QD = 0$ and π for the Bloch wave vector, which is similar to the situations described for superlattice polaritons in chapter 5.

6.3 Ferroelectric materials

Ferroelectrics are materials that undergo a phase transition below a critical temperature to a state with a spontaneous electric-dipole moment. Some general references on ferroelectricity are the books by Blinc and Zeks [33] and Lines and Glass [34]. As a prelude to tackling the problem of the modes or excitations in ferroelectrics, it is necessary to have an understanding of the static properties (i.e., modifications to the equilibrium electric-dipole moments) in the vicinity of a surface. This was notably studied by Kretschmer and Binder [35] using Landau theory (see below).

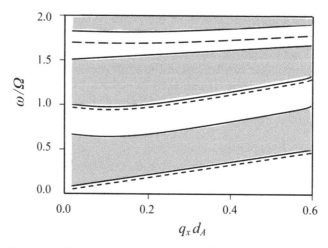

Figure 6.3. Dispersion relations for piezoelectric waves in a semi-infinite periodic superlattice with alternating layers of A (a non-piezoelectric material taken as fused silica with thickness $d_A = 20$ nm) and B (a hexagonal piezoelectric material taken as ZnO with thickness $d_B = 40$ nm). The bulk and surface modes are shown as the shaded areas and the dashed lines, respectively. The characteristic frequency Ω denotes v_{TA}/d_A where v_{TA} is the TA wave velocity in material A. After Albuquerque and Cottam [32].

The dynamical properties in the bulk and at surfaces of ferroelectrics will be discussed here using three different approaches, as in the treatment for surface excitations at a single interface in ferroelectrics made by Cottam, Tilley and Zeks [36]. One study is based on formal analogies with the transverse Ising model of ferromagnetism and is generally assumed to be appropriate for hydrogen-bonded ferroelectrics such as potassium dihydrogen phosphate (KDP). Another approach is an adaptation of the Landau theory of phase transitions where an expansion of the free energy is made in terms of an order parameter. Finally a theory can be developed using a polariton-type formalism (as in chapter 5) where now the dielectric function is given an appropriate anisotropic and temperature-dependent matrix form. Each of these methods will be described in turn.

6.3.1 The Ising model in a transverse field

The Ising model is well known as an exchange model of ferromagnetism in which the isotropic $\mathbf{S}_i \cdot \mathbf{S}_j$ exchange interaction of the Heisenberg model, as in equation (3.3), is replaced by the anisotropic product $S_i^z S_j^z$ where the preferred magnetization direction is along the z-axis. If there is a transverse applied magnetic field in the x-direction the Hamiltonian becomes

$$H = -\frac{1}{2}\sum_{i,j} J_{ij} S_i^z S_j^z - \sum_i \Omega_i S_i^z, \qquad (6.18)$$

where the notation is as before but with Ω_i proportional to the strength of the transverse field. The applicability of this Hamiltonian as a pseudo-spin model for hydrogen-bonded ferroelectrics is much discussed in the literature (see, e.g., [33, 34]).

In these materials a spontaneous electric-dipole moment occurs when there is ordering associated with the protons in the hydrogen bonds. Each proton has two possible positions, so it can be considered as being in a double potential well and mathematically represented by the eigenstates $S^z = \pm 1/2$ of an effective spin $S = 1/2$. The transverse-field term represents the ability of the proton to tunnel between the two states.

In an infinitely extended system described by equation (6.18) a phase transition is predicted at a temperature T_C given in mean-field (see [33]) by

$$\tanh(\Omega/2k_B T_C) = 2\Omega/n_0 J. \tag{6.19}$$

Here $\Omega \equiv \Omega_i$ at all sites and J denotes the exchange coupling to n_0 nearest neighbours. It can be shown that for temperature $T < T_C$ the static spin average at any site has components R^x and R^z in the x- and z-directions, respectively, but for $T \geqslant T_C$ only R^x is nonzero. The behaviour is sketched in figure 6.4. In fact, R^x is constant below T_C with the value $\Omega/n_0 J$ and then it monotonically decreases at higher temperatures. In the context of the transverse Ising model it is R^z that behaves as the order parameter for ferroelectricity, being proportional to the polarization. The bulk modes (essentially the magnons or SWs) corresponding to equation (6.18) for an infinite system were calculated in [33, 38].

Now, using methods similar to those described in subsection 3.2.1 for Heisenberg ferromagnets, we may deduce the surface modes in a semi-infinite ferroelectric. Thus, taking the coordinate axes as before and writing

$$\mathbf{S}_j = \mathbf{R}_j + \mathbf{s}_j, \tag{6.20}$$

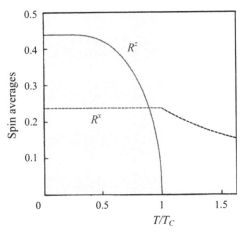

Figure 6.4. Spin averages R^x and R^z, calculated in mean-field theory for the Ising model in a transverse field, as a function of T/T_C. It has been assumed that $\Omega/n_0 J = 0.24$. After Cottam and Tilley [37].

we may employ the torque equation of motion as in (3.18) and a functional derivative for the effective field as in equation (3.21). A set of three linearized equations is then obtained for the components of $\mathbf{s}_j$ as

$$-i\omega s_j^x = s_j^x \sum_i J_{ij} R_i^z, \qquad -i\omega s_j^z = \Omega_j s_j^y,$$

$$-i\omega s_j^y = \Omega_j s_j^z - R_j^x \sum_i J_{ij} s_i^z - s_j^x \sum_i J_{ij} R_i^z, \qquad (6.21)$$

where quantities such as $\mathbf{R}_j$ and Ω_j may now depend on site label j. The solution of these equations is complicated in general [36], but here we consider for simplicity the solution in the high-temperature phase where $T > T_C$, allowing us to put $R_j^z = 0$ and $R_j^x = (1/2)\tanh(\Omega_j/2k_B T)$ according to mean-field theory. The equations in (6.21) then combine to give

$$-\left(\omega^2 - \Omega_j^2\right)s_j^z + \Omega_j R_j^x \sum_i J_{ij} s_i^z = 0. \qquad (6.22)$$

The solution of the above equation is now straightforward. We assume that the exchange J_{ij} couples only nearest neighbours, with the value J_S if both spins are in the surface and J otherwise (by analogy with section 3.2). Also we take Ω_j to be Ω_S if j is in the surface and Ω otherwise. The latter assumption means that R_j^x has only two values, which we denote as R_S^x if j is in the surface and R^x otherwise. If we now seek solutions for s_j^z with a dependence such as $s_n \exp(i\mathbf{q}_\parallel \cdot \mathbf{r}_\parallel - i\omega t)$, where the notation is the same as in section 3.2 with n ($= 1, 2, 3,...$) being the layer index, we have from equation (6.22)

$$\left\{\omega^2 - \Omega_S^2 + 4\Omega_S R_S^x J_S \gamma\left(\mathbf{q}_\parallel\right)\right\} s_1 + \Omega_S R_S^x J s_2 = 0,$$

$$\left\{\omega^2 - \Omega^2 + 4\Omega R^x J \gamma\left(\mathbf{q}_\parallel\right)\right\} s_n + \Omega R^x J(s_{n-1} + s_{n+1}) = 0 \quad (n > 1). \qquad (6.23)$$

The definition of $\gamma(\mathbf{q}_\parallel)$ is given in equation (3.10). The above equations are formally similar to those encountered in chapters 2 and 3 and can be solved by the same methods, i.e., either by seeking travelling-wave and attenuated solutions or by the TDM method described in the appendix. Here we just quote the results.

The bulk modes correspond to $\omega = \pm\omega_B(\mathbf{q})$ with the dispersion relation

$$\omega_B(\mathbf{q}) = \left[\Omega^2 - 2\Omega R^x J\left\{2\gamma(\mathbf{q}_\parallel) + \cos(q_z a)\right\}\right]^{1/2} \qquad (6.24)$$

assuming an sc crystal structure. The surface-mode solutions are more complicated, but in the special case of $\Omega_S = \Omega$ we have $\omega = \pm\omega_S(\mathbf{q}_\parallel)$ where

$$\omega_B(\mathbf{q}) = \left[\Omega^2 - \Omega R^x\left(4 J_S \gamma\left(\mathbf{q}_\parallel\right) + \frac{J}{4\sigma\gamma\left(\mathbf{q}_\parallel\right)}\right)\right]^{1/2}. \qquad (6.25)$$

Here $\sigma = (J_S - J)/J$, and equation (6.25) is applicable only if $J_S < (3/4)J$ or $J_S > (5/4)J$ as localization requirements. The bulk and surface dispersion relations are illustrated with numerical examples in figure 6.5.

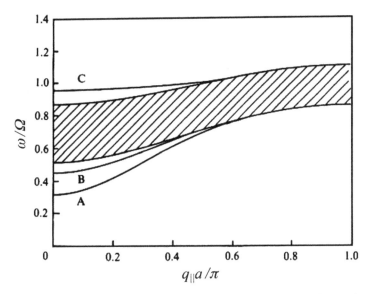

Figure 6.5. Reduced frequency ω/Ω plotted against $q_\parallel a/\pi$ for bulk (shaded region) and surface (lines) excitations for the Ising model in a transverse field for $\mathbf{q}_\parallel = (q_\parallel, 0)$. We have assumed $T = 1.5T_C$ and $\Omega_S = \Omega = 2J$. The J_S/J values for the three surface branches are: 'A' 1.75, 'B' 1.5 and 'C' 0.25. After Cottam *et al* [36].

In later work the analysis of ferroelectrics on the basis of the Ising model in a transverse field was extended to films (see [39]) and then to two-component periodic superlattices where both materials could be ferroelectrics (see, e.g., [40, 41]).

6.3.2 Landau theory for ferroelectrics

The basic idea in the Landau theory of phase transitions, e.g., as discussed in the books by Landau and Lifshitz [42], and Plischke and Bergersen [43], is that in the vicinity of the critical point the free energy of the system can be expanded in a power series involving the order parameter.

As in [36] we consider a simple ferroelectric material filling the half-space $z \geqslant 0$ and with just one component P of the polarization as the order parameter. It is assumed that the free energy F per unit area (in the xy-plane) has an expansion of the form

$$F = \int_0^\infty dz \left[\frac{1}{2}AP^2 + \frac{1}{4}BP^4 + \frac{1}{2}C(\nabla P)^2 - E_{ext}P \right] + \frac{1}{2}\xi CP^2(z = 0), \quad (6.26)$$

where A, B and C are the usual phenomenological expansion parameters used in this type of theory. They depend on temperature in general and A has the required properties that it vanishes at the critical temperature, so $A(T_C) = 0$, and $A(T) < 0$ for $T < T_C$. E_{ext} is an applied static electric field that couples linearly to the polarization. The final term in equation (6.26), with ξ denoting a constant, is introduced as a phenomenological boundary condition at $z = 0$.

The equilibrium polarization is found by applying the variational equation $\delta F/\delta P = 0$, giving

$$AP + BP^3 - C\nabla^2 P = E_{\text{ext}} \tag{6.27}$$

with the boundary condition being

$$\frac{dP}{dz} - \xi P = 0, \quad (z = 0). \tag{6.28}$$

We note that equation (6.28) has the same form as the exchange boundary condition in subsection 3.3.2. The effects of ξ are broadly analogous to the roles played by J_S/J and Ω_S/Ω in the Ising-model approach.

The mode frequencies for fluctuations in the polarization P can be investigated by postulating equations of motion in a damped oscillator form (see [36, 44]) as

$$m\frac{\partial^2 P}{\partial t^2} + \gamma\frac{\partial P}{\partial t} + \frac{\delta F}{\delta P} = 0, \tag{6.29}$$

where m is an inertial factor, γ is a damping factor and the next term is an effective force. Next we assume a time dependence as $\exp(-i\omega t)$ and linearize the equation of motion by writing $P = P_{\text{eq}} + p$, where P_{eq} is the equilibrium polarization. The result is

$$-Cd^2p\big/dz^2 + \left(-m\omega^2 - i\gamma\omega + A + 3BP_{\text{eq}}^2 + Cq_{\parallel}^2\right)p = 0, \tag{6.30}$$

where $q_{\parallel}$ is the wave vector parallel to the surface. A useful mathematical device to solve this equation is to define an eigenvalue E corresponding to the linear differential equation $-Cd^2p/dz^2 + 3BP_{\text{eq}}^2\,p = Ep$, whereupon the dispersion relation can be expressed formally as

$$m\omega^2 + i\gamma\omega = A + E + Cq_{\parallel}^2. \tag{6.31}$$

In solving for E and thus for the dependence of the mode frequency ω on $q_{\parallel}$ or T, formal results due to Lubensky and Rubin [45] may be used. To illustrate the predictions of the theory we take $A = A_0[(T/T_C) - 1]$ while B and C are independent of T. A numerical example of the temperature dependence for the surface mode in the case of $q_{\parallel} = 0$ is shown in figure 6.6 for $T < T_C$ and $\xi > 0$. In this case the surface mode, which goes 'soft' (tends to zero frequency) at T_C, lies below the bulk continuum region. Here the surface mode is well split off from the bottom of the bulk band, but if ξ is larger it moves closer. When $\xi < 0$ a modified surface mode can still exist under some conditions [36].

The type of approach described above for a semi-infinite ferroelectric can readily be generalized to apply to thin films and to alternating periodic superlattices composed of two different ferroelectric materials. Some useful references to specific calculations and applications are [46–49], while comprehensive reviews have been given by Ishibashi [50] and Chew [51] including applications to many different materials.

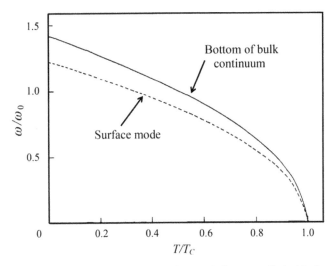

Figure 6.6. Predicted temperature dependence for the surface-mode frequency (dashed line) at $q_\| = 0$ for a semi-infinite ferroelectric material obtained using Landau theory. For comparison the bottom of the bulk-mode band corresponds to the solid line. It is assumed that $\xi = 0.1(A_0/C)^{1/2}$ and $\omega_0 = (A_0/m)^{1/2}$. After Cottam *et al* [36].

6.3.3 Polariton theory for ferroelectrics

In the electromagnetic regime of small wave vectors, such that $q \sim \omega/c$, we may follow a polariton description for the ferroelectric excitations. This has been applied both to the hydrogen-bonded ferroelectrics (such as KDP) and to displacive ferroelectrics (such as BaTiO$_3$).

Specifically, the polariton approach involves using an *anisotropic* form for the effective dielectric function written as a uniaxial matrix

$$\bar{\varepsilon} = \begin{pmatrix} \varepsilon_\perp & 0 & 0 \\ 0 & \varepsilon_\perp & 0 \\ 0 & 0 & \varepsilon_\| \end{pmatrix}, \tag{6.32}$$

where only the component $\varepsilon_\|$ displays a significant dependence on the ω and T characteristics of a ferroelectric [36]. In contrast, in most of the examples for non-magnetic polaritons in chapter 5 we employed an isotropic dielectric function. The exceptions were for magnetoplasmon-polaritons and where the effective-medium approximation was used in certain superlattices. It is the case, however, that the theory for polaritons in anisotropic media is well established, mainly due to work by Borstel and Falge [52, 53], using a more general matrix form than given in equation (6.32). If we restrict attention to TM modes and assume that the principal axes for the ferroelectric are aligned parallel and perpendicular to the surface, we find that the previous calculation in subsection 5.1.1 for the surface polariton dispersion relation in the isotropic case now leads to

$$q_x^2 = \frac{\omega^2}{c^2} \left(\frac{\varepsilon_A \varepsilon_\perp [\varepsilon_A - \varepsilon_\|]}{\varepsilon_A^2 - \varepsilon_\| \varepsilon_\perp} \right). \tag{6.33}$$

Here A denotes the inactive medium with isotropic dielectric constant ε_A in the half-space $z > 0$, while the ferroelectric (medium B) fills $z < 0$. We note that the above expression is a generalization of equation (5.6), which is recovered when we set $\varepsilon_\parallel = \varepsilon_\perp = \varepsilon_B$.

A suitable choice to be adopted for $\varepsilon_\parallel$ in the case of a ferroelectric material takes the same form as in equation (4.35) for optic phonons in an ionic crystal. In the present context we make the replacement $\omega_L \to \omega_0$ for the ferroelectric mode. In the absence of damping ($\Gamma = 0$) the frequency ω_0 is assumed to have the soft-mode property $\omega_0 \to 0$ as $T \to T_C$. At temperatures sufficiently close to T_C, however, the damping cannot be ignored and so either the full form of equation (4.35) is employed or a relaxational form such as

$$\varepsilon_\parallel(\omega) = \frac{\varepsilon_\parallel(0)}{1 + i\omega\tau} \tag{6.34}$$

may be appropriate, where τ denotes a relaxation time.

The existence conditions for surface polaritons can now be deduced by following the same steps as in chapter 5, i.e., by considering the signs of $\text{Im}(q_{Az})$ and $\text{Im}(q_{Bz})$ for localization. If we employ the modified equation (4.35), putting $\Gamma = 0$ and adapting results from [52, 53] for anisotropic materials, we find that necessary requirements are

$$(\varepsilon_A\omega^2/c^2) - q_x^2 < 0 \qquad \text{and} \qquad \varepsilon_\parallel\varepsilon_\perp\left[(\varepsilon_\perp\omega^2/c^2) - q_x^2\right] < 0. \tag{6.35}$$

In analysing the predictions of the above theory it is useful, when using equation (4.35) for $\varepsilon_\parallel$ with $\Gamma = 0$, to define a characteristic frequency ω_S by $\varepsilon_\parallel(\omega_S) = \varepsilon_A^2/\varepsilon_\perp$ and to recall the property that $\varepsilon_\parallel(\omega_L) = 0$. It can then be deduced that the surface polariton has the properties

$$\begin{aligned} q_x &= \varepsilon_A^{1/2}\omega/c \quad \text{when } \omega = \omega_0, \\ q_x &= \varepsilon_\perp^{1/2}\omega/c \quad \text{when } \omega = \omega_L, \end{aligned} \tag{6.36}$$

and $\omega \to \omega_S$ as $q_x \to \infty$. When the damping is nonzero but small, these surface-polariton results are modified and there is also some mixing with another mode that would have a purely bulk-like character when $\Gamma = 0$.

A numerical example of the surface-polariton dispersion relation is shown in figure 6.7 assuming relatively weak damping with $\Gamma/\omega_0 = 0.2$. Here ω is taken as real and so q_x is complex in the presence of damping (which makes the dielectric function $\varepsilon_\parallel$ complex). The real and imaginary parts of q_x are plotted for real ω, which is the representation that would be relevant for an ATR scan (see subsection 5.1.2). In this context the imaginary part $\text{Im}(q_x)$ is interpreted as the reciprocal of the decay length for the surface polariton. The main features in the behaviour for both $\text{Re}(q_x)$ and $\text{Im}(q_x)$ are resonances around $\omega \simeq \omega_S$. Polariton calculations for the heavily damped regime, obtained using equation (6.34) for $\varepsilon_\parallel$, are given in [36] and are typically non-resonant in form.

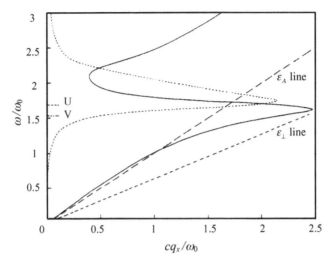

Figure 6.7. Dispersion properties of a TM surface polariton on an anisotropic medium representing a ferroelectric, as calculated using equation (6.33). The mode frequency is plotted against $Re(q_x)$ (solid curves) and $Im(q_x)$ (broken curves). The straight lines labelled as 'ε_A line' and '$\varepsilon_\perp$ line' refer to the conditions in equation (6.36). The frequency intercepts 'U' and 'V' refer to ω_S and ω_L, respectively. The parameters are $\Gamma/\omega_0 = 0.2$, $\varepsilon_A = 1$, $\varepsilon_\perp = 2.5$, $\varepsilon_\infty = 1.5$ and $\omega_L/\omega_0 = 1.53$. After Cottam *et al* [36].

6.4 Multiferroic materials

In this last section we provide a brief account of multiferroics, which can be defined (see, e.g., [54]) as materials that can simultaneously in the same phase exhibit at least two of the following types of behaviour: *ferromagnetism*, *ferroelectricity* and *ferroelasticity*. On the magnetic side, more complicated spin-ordering schemes as found in antiferromagnetism and ferrimagnetism are also included in the definition. We have already discussed ferromagnetic and ferroelectric properties individually in this book, whereas the third category of ferroelastic materials may broadly be defined as those which display a spontaneous strain below a phase transition (by analogy with the spontaneous magnetization or polarization in the other cases).

To date it has been the simultaneous occurrence of magnetic and ferroelectric behaviour that has attracted the most interest and therefore these types of multiferroics (sometimes called *magnetoelectric* multiferroics) will be the focus of our discussion here. Typically the problem is in achieving a magnetoelectric coupling between the coexisting magnetization **M** and polarization **P**, so that the magnetic excitations may in principle be controlled by an electric field or conversely the polarization dynamics by a magnetic field.

The early work on dynamical processes in multiferroics (in the 1960s to 1980s) notably included calculations by Bar'yakhtar and Chupis [55] and Maugin [56] for the role of the magnetoelectric coupling in modifying the magnon frequencies in ferroelectric ferromagnets. In effect, they studied electric-dipole-active magnons, sometimes referred to as *electromagnons*. The methods employed were analogous to those described for magnons in chapter 3 but with an extra coupling term that was quadratic in both the polarization and the magnetization components. The main difference in the two calculations was that Maugin included magnetostatic effects that were ignored in the previous paper. At around the same time Tilley and Scott [57] used a Landau-type theory to study the high-frequency modes in the antiferromagnet $BaMnF_4$ with an assumed magnetoelectric coupling. This material is a two-sublattice antiferromagnet with a complicated phase behaviour whereby the sublattice magnetization directions are canted slightly (i.e., they are not exactly at 180° to one another). Hence $BaMnF_4$ behaves as a *weak ferromagnet* and there can be a low-frequency magnon associated with precession about this weak magnetic moment, as well as the higher-frequency magnon branch typical of an antiferromagnet.

Since the predicted, as well as observed, effects of the magnetoelectric coupling were so small, interest in multiferroics declined. It is only fairly recently in the last decade that the discovery of new materials with more complicated magnetic and electric ordering has led to a great revival in this field. In particular, we mention that an issue of *Nature Materials* (volume 6, number 1) in 2007 was devoted to this topic. It included review articles by Cheong and Mostovoy [58], and Ramesh and Spaldin [59] detailing some of the advances for thin-film multiferroics and the potential applications for electronic devices and sensors.

Corresponding developments on the theoretical side have also been spurred on and we now outline as an example some studies by de Sousa and Moore [60]. As in [57] they considered a weak ferromagnetic (canted antiferromagnetic) material where the polarization is perpendicular to the weak ferromagnetic moment. The basic geometry and choice of coordinate axes are as depicted in figure 6.8, taking the equilibrium polarization to be perpendicular to the plane of the sublattice magnetization vectors (and hence also the weak ferromagnetic moment). The model is appropriate to multiferroics such as $BiFeO_3$ where the canting angle θ is of order 0.1°.

The free energy per unit volume for the system is assumed, following [60], to have the form $F = F_E + F_M + F_{ME}$, where F_E comprises the Landau expansion terms (analogous to those in equation (6.26) but omitting the surface term for simplicity) for the ferroelectric properties in terms of **P**. Next F_M involves similar expansions for the magnetic system in terms of the sublattice magnetizations $\mathbf{M_1}$ and $\mathbf{M_2}$. The coefficients are such as to give the z-axis as the preferred direction for **P** and the

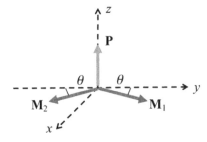

Figure 6.8. Geometry for a ferroelectric canted antiferromagnet (weak ferromagnet). The equilibrium polarization is along the z-direction, while the equilibrium sublattice magnetizations are in the xy-plane with a small canting angle θ away from the y-axis. The weak ferromagnetic moment is along the x-direction.

$\pm y$-axes as the preferred directions for M_1 and M_2, respectively. Finally, the magnetoelectric contribution is

$$F_{ME} = \left(J + \alpha P^2\right)M_1 \cdot M_2 + \beta P \cdot (M_1 \times M_2), \qquad (6.37)$$

where α and β are coupling constants. The first term on the right-hand side describes the nearest-neighbour exchange between opposite sublattices, by analogy with results in subsection 3.2.2, except that now a term proportional to P^2 is introduced into the effective exchange $(J + \alpha P^2)$. The second term, being proportional to the combination $M_1 \times M_2$, is an example of a Dzyaloshinskii–Moriya (DM) interaction [61, 62] and is allowed by symmetry in BiFeO$_3$. It provides the mechanism for the canting of the sublattice magnetizations and here its strength is proportional to the polarization. Minimization of the free energy leads to the result for the canting angle θ that [60]

$$\tan \theta = \frac{\beta P_0}{\left(J + \alpha P_0^2\right) + \left[\left(\beta P_0\right)^2 + \left(J + \alpha P_0^2\right)^2\right]^{1/2}}, \qquad (6.38)$$

where P_0 is the magnitude of the equilibrium polarization.

The dynamics for the coupled fluctuations in the magnetization vectors and in the polarization were studied by de Sousa and Moore [60] by using the Landau–Lifshitz torque equation of motion, by analogy with equation (3.17), and Maxwell's equations of electromagnetism, in the magnetostatic approximation, respectively. They found, on examining a limit where fluctuations in the polarization were small, that there are two SW branches (or electromagnons) each with a dependence on P_0. The low-frequency SW mode corresponds to spin fluctuations that leave the overall canting angle unchanged, since the spins precess about their equilibrium directions in phase. For a propagation wave vector q in the xz-plane this mode is gapless in the sense that its frequency is proportional to q. In contrast, for propagation along the y-axis (the direction of the weak ferromagnetic moment) the frequency is nonzero and proportional to βP_0 as $q \rightarrow 0$. This gap arises due to a magnetostatic correction in the presence of the DM interaction. The high-frequency SW mode is less interesting; it changes the overall canting angle but otherwise is broadly similar to the usual antiferromagnetic SW in a simple antiferromagnet. A fuller description of these modes, including their q-dependence and the magnetic polariton regime, can be found in [60]. The experimental detection of electromagnons in BiFeO$_3$ was reported in [63].

Further calculations for the electromagnons, using broadly similar methods and again with a model of a DM canted antiferromagnet, were presented by Livesey and Stamps [64]. They focused on obtaining the dynamic susceptibility over a wide frequency range, including that corresponding to the high-frequency branch for the electromagnons. They also used an effective-medium approach applied to a hetero-structure superlattice of alternating layers of BiFeO$_3$ (the multiferroic component) and permalloy (the ferromagnetic component). Experimental work on other types of multiferroic superlattices has been reported, for example, by Yang *et al* [65].

Finally we mention that there have been recent studies of multiferroic behaviour in other low-dimensional systems, in particular for cylindrical nanowire and nanotube geometries by Eliseev *et al* [66]. They considered $Eu_xSr_{1-x}TiO_3$ for various concentrations x of Eu and calculated the phase diagrams under various conditions of stress in these low-dimensional geometries.

References

[1] Turov E A and Irkhin Y P 1956 *Fiz. Met. Metalloved.* **3** 15
[2] Kittel C 1958 *Phys. Rev.* **110** 836
[3] Schlömann E 1960 *J. Appl. Phys.* **31** 1647
[4] Bland J A C and Heinrich H (ed) 1994 *Ultrathin Magnetic Structures* vol 1 (Berlin: Springer)
[5] Akhiezer A I, Bar'yakhtar V G and Peletminskii S V 1968 *Spin Waves* (Amsterdam: North-Holland)
[6] Keffer F 1966 *Handbuch der Physik* vol 18 (Berlin: Springer) p 1
[7] L'vov V S 1991 *Wave Turbulence under Parametric Excitation* (Berlin: Springer)
[8] Gurevich A G and Melkov G A 1996 *Magnetization Oscillations and Waves* (Boca Raton, FL: CRC)
[9] Landau L D and Lifshitz E M 1970 *Theory of Elasticity* (Oxford: Pergamon)
[10] Scott R Q and Mills D L 1977 *Phys. Rev.* B **15** 3545
[11] Temiryazev A G, Tikhomirova M P and Zilberman P E 1996 *Nonlinear Microwave Signal Processing: Towards a New Range of Devices* ed M Marcelli and S A Nikitov (Dordrecht: Kluwer) p 165
[12] Gulyaev Yu V and Zil'berman P E 1988 *Russ. Phys. J* **31** 860
[13] Gorobets Yu I, Reshetnyal S A and Khomenko T A 2010 *Acta Phys. Pol.* A **117** 211
[14] Tarasenko O S, Tarasenko S V and Yurchenko V M 2007 *Crystallogr. Rep.* **51** 296
[15] Belykh N V, Lapteva T V, Tarasenko O S and Tarasenko S V 2007 *Crystallogr. Rep.* **52** 864
[16] Lingner C and Lüthi B 1981 *Phys. Rev.* B **23** 256
[17] Royer D and Dieulesaint E 2000 *Elastic Waves in Solids* (Berlin: Springer) vol 1 and 2
[18] Manbachi A and Cobbold R S C 2011 *Ultrasound* **19** 187
[19] Cady W G 1946 *Piezoelectricity* (New York: McGraw-Hill)
[20] Maradudin A A 1981 *Advances in Solid State Physics* vol 21 (Berlin: Springer) p 25
[21] Cottam M G and Maradudin A A 1984 *Surface Excitations* ed V M Agranovich and R Loudon (Amsterdam: North-Holland) p 1
[22] Albuquerque E A and Cottam M G 2004 *Polaritons in Periodic and Quasiperiodic Structures* (Amsterdam: Elsevier)
[23] Tseng C C 1967 *J. Appl. Phys.* **38** 4281
[24] Bleustein J L 1968 *Appl. Phys. Lett.* **13** 412
[25] Gulyaev Y V 1969 *JETP Lett.* **9** 37
[26] Albuquerque E L 1979 *Phys. Status Solidi* B **96** 475
[27] Albuquerque E L 1981 *Phys. Status Solidi* B **104** 667
[28] Auld B A 1990 *Acoustic Fields and Waves in Solids* vol 1 and 2 2nd edn (Malabar, FL: Krieger)
[29] Wherrett B S 1986 *Group Theory for Atoms, Molecules and Solids* (Englewood Cliffs, NJ: Prentice-Hall)
[30] Akcakaya E and Farnell G W 1988 *J. Appl. Phys.* **64** 4469
[31] Akcakaya E, Farnell G W and Adler E L 1990 *J. Appl. Phys.* **64** 4469

[32] Albuquerque E A and Cottam M G 1992 *Solid State Commun.* **83** 545
[33] Blinc R and Zeks B 1974 *Soft Modes in Ferroelectrics and Antiferroelectrics* (Amsterdam: North-Holland)
[34] Lines M E and Glass A M 1979 *Principles and Applications of Ferroelectrics and Related Materials* (Oxford: Oxford University Press)
[35] Kretschmer R and Binder K 1979 *Phys. Rev.* B **20** 1065
[36] Cottam M G, Tilley D R and Zeks B 1984 *J. Phys. C: Solid State Phys.* **17** 1793
[37] Cottam M G and Tilley D R 2005 *Introduction to Surface and Superlattice Excitations* 2nd edn (Bristol: IOP)
[38] Stinchcombe R B 1973 *J. Phys. C: Solid State Phys.* **6** 2459
[39] Salman A and Cottam M G 1994 *Surf. Rev. Lett.* **1** 23
[40] El Aouad N, Laaboudi B, Kerouad M and Saber M 2001 *J. Phys.: Condens. Matter* **13** 797
[41] Saber M, Ainane A and Bärner K 2007 *Surf. Sci.* **601** 4274
[42] Landau L D and Lifshitz E M 1969 *Statistical Physics* (Oxford: Pergamon)
[43] Plischke M and Bergersen B 1989 *Equilibrium Statistical Physics* (Englewood, NJ: Prentice-Hall)
[44] Ginzburg V L, Levanyuk A P and Sobyanin A A 1980 *Phys. Rep.* **57** 151
[45] Lubensky T C and Rubin M H 1975 *Phys. Rev.* B **12** 3885
[46] Tilley D R and Zeks B 1984 *Solid State Commun.* **49** 823
[47] Schwenk D, Fishman F and Schwabl F 1990 *Ferroelectrics* **104** 349
[48] Li S, Eastman J A, Vetrone J M, Newnham R E and Cross L E 1997 *Phil. Mag.* B **76** 47
[49] Ong L-H, Osman J and Tilley D R 2001 *Phys. Rev.* B **63** 144109
[50] Ishibashi Y 2005 *Ferroelectric Thin Films: Basic Properties and Device Physics for Memory Applications* ed M Okuyama and Y Ishibashi (Berlin: Springer)
[51] Chew K H 2012 *Solid State Phenom.* **189** 145
[52] Borstel G and Falge H J 1977 *Phys. Status Solidi* B **83** 11
[53] Borstel G and Falge H J 1978 *J. Appl. Phys.* **16** 211
[54] Schmid H 1994 *Ferroelectrics* **162** 317
[55] Bar'yakhtar V G and Chupis I E 1969 *Sov. Phys.—Solid State* **10** 2818
     Bar'yakhtar V G and Chupis I E 1970 *Sov. Phys.—Solid State* **11** 2628
[56] Maugin G A 1981 *Phys. Rev.* B **23** 4608
[57] Tilley D R and Scott J F 1982 *Phys. Rev.* B **25** 3251
[58] Cheong S-K and Mostovoy M 2007 *Nat. Mater.* **6** 13
[59] Ramesh R and Spaldin N A 2007 *Nat. Mater.* **6** 21
[60] de Sousa R and Moore J E 2008 *Appl. Phys. Lett.* **92** 022514
[61] Dzyaloshinskii I E 1958 *J. Phys. Chem. Solids* **4** 241
[62] Moriya T 1960 *Phys. Rev.* **120** 9
[63] Cazayous M, Gallais Y, Sacuto A, de Sousa R, Lebeugle D and Colson D 2008 *Phys. Rev. Lett.* **101** 037601
[64] Livesey K L and Stamps R L 2010 *Phys. Rev.* B **81** 094405
[65] Yang Z, Ke C, Sun L L, Zhu W, Wang L, Chen X F and Tan O K 2010 *Solid State Commun.* **150** 1432
[66] Eliseev A A, Glinchuk M D, Khist V V, Lee C-W, Deo C S, Behera R K and Morozovska A N 2013 *J. Appl. Phys.* **113** 024107

IOP Publishing

Dynamical Properties in Nanostructured and
Low-Dimensional Materials

Michael G Cottam

Chapter 7

Nonlinear dynamics of excitations

The preceding chapters have been mainly devoted to the linear aspects of the dynamics, which have been the subject of detailed studies. We introduced, however, the concept of NL excitations in chapter 1 (see subsection 1.1.2) where we described NL aspects of the dielectric function and outlined a calculation for a NL surface wave that had no counterpart in the linear regime. Later in chapter 3 we briefly indicated how NL effects might arise for SWs or magnons in magnetic systems in terms of boson operator methods; we refer to equation (3.73) and the following text.

The aim of this chapter is to develop in particular cases the NL dynamical properties for the excitations covered earlier in the book, especially for low-dimensional structures and for surface–interface geometries. This is not intended as a comprehensive treatment of the NL dynamics, since that is a major topic in its own right and there are already textbooks dealing with specialized aspects. Instead it is included as a concise survey with examples chosen to relate to the material of the previous chapters and to recent developments. As a general reference the book by Cottam and Tilley [1] includes three chapters on NL waves summarizing work in this field up to about 2005.

7.1 Fundamentals of optical and magnetic nonlinearities

We start by summarizing here some of the basic results, including the origins of the nonlinearities and the different approaches that may be followed for the excitations in either calculating their properties or probing them experimentally. It will be convenient to consider separately the approach conventionally followed in NL optics (and other areas of electromagnetism) from that in magnetic systems. In the former case the development is typically in terms of an expansion of the NL dielectric function, or its corresponding susceptibility, as already discussed briefly in subsection 1.1.2. In the latter case, depending on the magnetic regime for the

doi:10.1088/978-0-7503-1054-3ch7

excitation wave vector, there is either a macroscopic method in terms of a NL magnetic susceptibility (the analogue of the NL optics approach) or a microscopic dipole-exchange method. Also in a later part of this section we introduce solitons (and solitary waves), which have no counterpart in the the linear regime.

7.1.1 Nonlinear waves in optics

Some general NL optics references are the books by Yariv [2], Butcher and Cotter [3], and Mills [4], which provide detailed accounts of the theoretical formalism. Already, in chapter 1, we discussed the response of a dielectric material to a position- and time-dependent electric field $\mathbf{E}$. The important relationships involving the polarization $\mathbf{P}$ and the dielectric function, $\varepsilon(\mathbf{q}, \omega)$ if spatial dispersion is included, are those given in equations (1.13)–(1.17). We may also add the definition of the susceptibility $\chi(\mathbf{q}, \omega)$ or $\chi(\omega)$, as $\mathbf{P}(\mathbf{q}, \omega) = \varepsilon_0 \chi(\mathbf{q}, \omega)\mathbf{E}(\mathbf{q}, \omega)$ and so from equation (1.17) we have the identity that

$$\varepsilon(\mathbf{q}, \omega) = 1 + \chi(\mathbf{q}, \omega). \tag{7.1}$$

The conventional approach in NL optics is now to expand the polarization, or equivalently the susceptibility or the dielectric function, in terms of Cartesian components of the electric field. Hence the expressions for the linear (L) and NL parts of the Fourier component of the polarization at any single frequency (say, ω_0), take the form

$$P_i^{\mathrm{L}}(\omega_0) = \sum_j \chi_{ij}^{(1)}(-\omega_0; \omega_1)E_j(\omega_1), \tag{7.2}$$

$$P_i^{\mathrm{NL}}(\omega_0) = \sum_{jk} \chi_{ijk}^{(2)}(-\omega_0; \omega_1, \omega_2)E_j(\omega_1)E_k(\omega_2)$$
$$+ \sum_{jkl} \chi_{ijlk}^{(3)}(-\omega_0; \omega_1, \omega_2, \omega_3)E_j(\omega_1)E_k(\omega_2)E_l(\omega_3) + \cdots. \tag{7.3}$$

On the right-hand side there are terms involving Fourier components of the electric-field vectors at the indicated frequencies ω_1, ω_2, etc. Each of these frequencies can be positive or negative (or zero). In equation (7.2) the linear susceptibility is just the second-order tensor (or matrix) with elements $\chi_{ij}^{(1)}$, which we have already discussed (in terms of the linear dielectric function) in chapter 4 and elsewhere. The NL susceptibility terms in equation (7.3) start with $\chi_{ijk}^{(2)}$ and $\chi_{ijkl}^{(3)}$, which are the elements of third- and fourth-order tensors, respectively. The allowed nonzero elements of all these susceptibility tensors can be deduced by symmetry arguments (see, e.g. [5, 6]) and these elements may be different at surfaces compared with the bulk of a material.

The notation used for the frequency dependence of the susceptibility terms follows that in [3], where the general term $\chi_{ij\cdots m}^{(n)}(-\omega_0; \omega_1, \omega_2, \ldots, \omega_n)$ has the property that

$$\omega_0 = \omega_1 + \omega_2 + \cdots + \omega_n. \tag{7.4}$$

In the linear case, no new frequencies are produced since we simply have $\omega_0 = \omega_1$. Various different situations arise, however, with the higher-order terms and in the following paragraphs we briefly consider the second- and third-order NL effects.

The possibilities for the overall frequency from the second-order $\chi^{(2)}$ nonlinearities are the sum and difference terms such as $|\omega_1| \pm |\omega_2|$. This can lead to a frequency doubling of the initial frequency in the sense that if the input frequencies of the electric fields are $\omega_1 = \omega_2 \equiv \omega$, then the output signal for the NL polarization involves a frequency $\omega_0 = 2\omega$. This is known as *second-harmonic generation* (SHG). Likewise, there will be a difference-frequency effect coming at $\omega_0 = 0$ and this is *optical rectification*. Another case of interest is when one of the input fields is static (e.g., with $\omega_1 = 0$) and the other has frequency $\omega_2 \equiv \omega$. The output frequency for the NL polarization is then also at ω, which is the same as in the linear effect but the total polarization now has an additional contribution related to $\chi_{ijk}^{(2)}(-\omega; 0, \omega)$. This is known as *Pockel's effect*.

Further NL effects arise in third order from the term $\chi_{ijlk}^{(3)}(-\omega_0; \omega_1, \omega_2, \omega_3)$ when various choices are made for the frequencies. In particular, we mention the cases of

$$\chi_{ijlk}^{(3)}(-3\omega; \omega, \omega, \omega) \qquad \text{and} \qquad \chi_{ijlk}^{(3)}(-\omega; \omega, -\omega, \omega). \qquad (7.5)$$

The first of these corresponds to *third-harmonic generation*. The second case produces output at the same input frequency ω and, together with similar terms, it gives rise to what is known as the *Kerr effect*. It follows that the total effective dielectric function at the frequency ω is then given by

$$\varepsilon_{ij}^{\text{eff}} = \varepsilon_{ij} + \frac{3}{4} \sum_{lk} \chi_{ijlk}^{(3)}(-\omega; \omega, -\omega, \omega) E_k(-\omega) E_l(\omega), \qquad (7.6)$$

where the factor of 3/4 comes from a careful consideration of the permutations involved. This simplifies if we assume an isotropic material (such as glass), since the linear term reduces to a scalar (multiplied by the unit matrix) and the only nonzero elements of the NL tensor are equivalent diagonal terms such as $\chi_{xxxx}^{(3)}$. Hence the modified scalar dielectric function is

$$\varepsilon^{\text{eff}} = \varepsilon + \frac{3}{4}\chi_{xxxx}^{(3)} |\mathbf{E}|^2, \qquad (7.7)$$

which is the same result as quoted in equation (1.18) with the NL coefficient given by $\rho = (3/4)\chi_{xxxx}^{(3)}$. We now see the origin of the intensity-dependent contribution to the dielectric function described in subsection 1.1.2 as being the Kerr effect. The field dependence predicted in equation (7.7) eventually breaks down at large enough $|\mathbf{E}|^2$ and there is a saturation effect whereby ε^{eff} tends to a large constant value.

Apart from the specific NL effects described above, the $\chi^{(2)}$ and $\chi^{(3)}$ terms can be used in many-body perturbation theory to calculate the contributions to the damping, which was introduced phenomenologically into many expressions for the dielectric functions in chapter 4.

7.1.2 Nonlinear waves in magnetic systems

In the treatments of magnons or SWs in chapter 3, approximations were introduced that made the equations of motion for any transverse component of the dynamic magnetization (or alternatively a spin operator) become linear in the amplitudes. A consequence was that fluctuations were neglected in the longitudinal z-component of magnetization or spin, where the z-axis defines the direction of the static magnetization in a ferromagnet. It is important to realize, however, that the starting equations (e.g., the Heisenberg spin Hamiltonian in equation (3.3) or the torque equation for the total magnetization in equation (3.17)) are actually NL. The role of magnetic nonlinearities becomes more important at elevated temperatures or in experiments at higher power levels (e.g., in FMR with high-intensity pumping magnetic fields). Some general references, mainly for NL properties of bulk magnons, are by Keffer [7], L'vov [8], Gurevich and Melkov [9], and Stancil and Prabhakar [10]. Some other books that emphasize surface and interface properties are found in [11–15].

Suppose we start with the torque equation of motion for the total magnetization $\mathbf{M}$ of a ferromagnet (3.17). It follows that $\mathbf{M} \cdot d\mathbf{M}/dt = 0$ and hence $d(\mathbf{M} \cdot \mathbf{M})/dt = 0$ implying that $|\mathbf{M}|$ is a constant, which we denote by M_0. Hence the components of the fluctuating magnetization $\mathbf{m}$ satisfy

$$m_x^2 + m_y^2 + (M_0 + m_z)^2 = M_0^2. \tag{7.8}$$

This can be rearranged to give

$$m_z \simeq -\frac{m_T^2}{2M_0}, \tag{7.9}$$

where $\mathbf{m}_T = (m_x, m_y)$ is the transverse magnetization and we have assumed $m_T \ll M_0$. The simple result in equation (7.9) not only provides the justification for neglecting m_z in a first approximation, since it is second order in the small quantity m_T, but it also points the way to developing an improved theory.

As was the case for the linear theory in sections 3.2 and 3.3, it is possible to formulate the results using either microscopic or macroscopic methods. We first describe the main NL processes in terms of three-magnon and four-magnon interactions. Then we include other (parametric) processes that occur when there is a 'pumping' electromagnetic field as in high-power FMR.

A. Three- and four-magnon processes

We may straightforwardly calculate the leading-order NL terms within a Hamiltonian formalism by extending the boson method used in subsection 3.3.2 for dipole-exchange waves. At that stage we introduced the HP transformation from spin operators to boson operators and a Hamiltonian expansion in equation (3.73) where the general $H^{(m)}$ term (with $m = 2, 3, 4,...$) contains products of m operators. Only the role of $H^{(2)}$ was discussed in chapter 3, since it accounts for the linearized SWs or magnons. In the context of a thin-film geometry $H^{(2)}$ was quoted in equation (3.74) in terms of the original boson operators (a and $a^\dagger$) and in a 'diagonalized' form in equation (3.75) after a linear transformation to new operators b and $b^\dagger$. The corresponding results for an infinite ferromagnet take an analogous but slightly simpler form (see, e.g., [7]) as

$$H^{(2)} = \sum_{\mathbf{q}} \left[A^{(2)}(\mathbf{q}) a^{\dagger}(\mathbf{q}) a(\mathbf{q}) + \left\{ B^{(2)}(\mathbf{q}) a^{\dagger}(\mathbf{q}) a^{\dagger}(-\mathbf{q}) + \text{H.c.} \right\} \right]$$

$$= \sum_{\mathbf{q}} \omega_B(\mathbf{q}) b^{\dagger}(\mathbf{q}) b(\mathbf{q}), \qquad (7.10)$$

where $\mathbf{q}$ is a 3D wave vector and there is only one magnon branch with the dispersion relation $\omega = \omega_B(\mathbf{q})$ and

$$\omega_B(\mathbf{q}) = \left\{ \left[A^{(2)}(\mathbf{q}) \right]^2 - |B^{(2)}(\mathbf{q})|^2 \right\}^{1/2}. \qquad (7.11)$$

Going now beyond the $H^{(2)}$ term to incorporate the NL dynamical effects, we find for an infinite ferromagnet that the next two terms in the expansion are

$$H^{(3)} = \sum_{\mathbf{q}_1, \mathbf{q}_2} \left[A^{(3)}(\mathbf{q}_1, \mathbf{q}_2) b^{\dagger}(\mathbf{q}_1) b^{\dagger}(\mathbf{q}_2) b(\mathbf{q}_1 + \mathbf{q}_2) + \text{H.c.} \right] \qquad (7.12)$$

$$H^{(4)} = \sum_{\mathbf{q}_1, \mathbf{q}_2, \mathbf{q}_3} \left[A^{(4)}(\mathbf{q}_1, \mathbf{q}_2, \mathbf{q}_3) b^{\dagger}(\mathbf{q}_1) b^{\dagger}(\mathbf{q}_2) b(\mathbf{q}_3) b(\mathbf{q}_1 + \mathbf{q}_2 - \mathbf{q}_3) \right.$$

$$\left. + \left\{ B^{(4)}(\mathbf{q}_1, \mathbf{q}_2, \mathbf{q}_3) b^{\dagger}(\mathbf{q}_1) b^{\dagger}(\mathbf{q}_2) b^{\dagger}(\mathbf{q}_3) b(\mathbf{q}_1 + \mathbf{q}_2 + \mathbf{q}_3) + \text{H.c.} \right\} \right]. \qquad (7.13)$$

Expressions for the amplitude functions $A^{(3)}$, $A^{(4)}$ and $B^{(4)}$ can be found, for example, in [7], where it is shown that when dipole–dipole interactions are neglected (i.e., in the Heisenberg limit) only $A^{(4)}$ is nonzero. A physical interpretation of the terms in the above equation can be made with the aid of figure 7.1, where each of the heavy lines represents propagation of a magnon. We follow the usual convention that an arrow pointing away from (towards) an interaction vertex indicates a creation (annihilation) operation. Figures 7.1(a) and (b) represent three-magnon *splitting* and *confluence* processes, respectively, with conservation of wave vector. Figure 7.1(c) shows an example of a four-magnon scattering process, again with conservation of wave vector. The role of energy conservation depends on the application, and examples will arise later.

The same formal results as described above can also be obtained in the macroscopic (or continuum) limit [7], but with some limited simplifications for the amplitudes. For example, the $A^{(2)}$ and $B^{(2)}$ terms of the linear theory can be expressed as

$$A^{(2)}(\mathbf{q}) = \omega_0 + Dq^2 + \frac{1}{2}\omega_M \sin^2\theta, \qquad B^{(2)}(\mathbf{q}) = \frac{1}{2}\omega_M \sin^2\theta \exp(2i\phi), \qquad (7.14)$$

where θ and ϕ are the polar and azimuthal angles for the 3D wave vector $\mathbf{q}$ in spherical polar coordinates and the other notation is the same as in chapter 3.

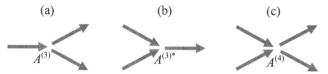

(a)   (b)   (c)

$A^{(3)}$ $A^{(3)*}$ $A^{(4)}$

Figure 7.1. Representation of the important interaction processes between magnons corresponding to terms in equations (7.12) and (7.13): (a) three-magnon splitting process, (b) three-magnon confluence process and (c) four-magnon scattering process.

B. Instability of magnons under microwave pumping

We now generalize the preceding paragraphs to incorporate the parametric instabilities that can occur for the magnons when an external magnetic pumping field is applied. This first attracted attention in some FMR experiments carried out at high power levels by Bloembergen and co-workers [16, 17]. Surprisingly they observed a saturation of the main FMR absorption peak and the appearance of subsidiary peaks, all with pronounced power thresholds. A theoretical interpretation was provided by Suhl [18] in terms of parametric processes. This stimulated related work and shortly afterwards Schlömann *et al* [19] demonstrated an analogous instability effect under parallel pumping with a microwave magnetic field. This was remarkable because the conventional FMR configuration is with the microwave field in the perpendicular orientation with respect to the magnetization direction.

The overall effect in the above processes is the decay of one (or more) uniform-precession SWs, meaning magnons with wave vector equal to zero, at the bulk frequency ω_B into a pair of magnons with frequencies ω_1 and ω_2 and corresponding wave vectors $\mathbf{q}_1$ and $\mathbf{q}_2$. The governing conservation conditions for an nth-order instability of this type are

$$\omega_1 + \omega_2 = n\omega_B, \qquad \mathbf{q}_1 + \mathbf{q}_2 = 0. \tag{7.15}$$

The simplest cases are for $n = 1$ and 2, which correspond to processes analysed in Suhl's original work.

Within our boson Hamiltonian approach we now add another term to represent the coupling of the magnetization components to a pumping field of the form $\mathbf{h}_p \exp[i\mathbf{q}_p \cdot \mathbf{r} - i\omega_p t]$. Often we may put the pumping wave vector $\mathbf{q}_p = 0$ for spatially homogeneous pumping in a typical FMR experiment. Apart from a constant term, the effective Hamiltonian has the form

$$H^{(2)} = \sum_{\mathbf{q}} \omega_B(\mathbf{q}) b^\dagger(\mathbf{q}) b(\mathbf{q}) + H^{(3)} + H^{(4)} + H_p, \tag{7.16}$$

where $H^{(3)}$ and $H^{(4)}$ are given by equations (7.12) and (7.13), respectively, and there are two cases for the pumping term which is proportional to the Zeeman energy $-\mathbf{h}_p \cdot \mathbf{m}$. For perpendicular pumping the coupling Hamiltonian $H_{p,\perp}$ is proportional to the transverse dynamic magnetization $\mathbf{m}_T$, whereas for parallel pumping $H_{p,\parallel}$ is proportional to the longitudinal dynamic magnetization m_z. Therefore they are distinct cases in terms of the boson operators involved in the pumping.

Perpendicular pumping occurs when there is a nonzero perpendicular (or transverse) component $h_{p,\perp}$ of the pumping field, resulting in

$$H_{p,\perp} \propto h_{p,\perp} \left[b^\dagger(0) \exp(-i\omega_p t) + \text{H.c.} \right], \tag{7.17}$$

to lowest order in the boson operators, assuming spatially homogeneous pumping. This just describes the usual FMR phenomenon with excitation of a uniform ($\mathbf{q} = 0$)

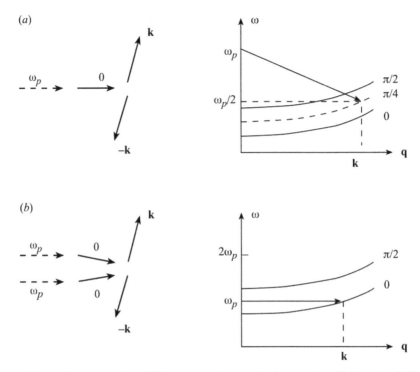

Figure 7.2. Schematic representation of NL processes in perpendicular pumping: (a) first-order Suhl process and (b) second-order Suhl process. On the left-hand side the solid and dashed lines refer to the magnon and electromagnetic modes, respectively. The dispersion curves are shown on the right-hand side. After Cottam and Tilley [1].

magnon. An analogy can be made with a damped harmonic oscillator (as in section 1.4), so that the mode amplitude becomes expressible in a Lorentzian form [18] as

$$|b(0)| = h_{p,\perp} \frac{U}{\left[(\omega_0 - \omega_p)^2 + \Gamma_0^2 \right]^{1/2}}. \tag{7.18}$$

Here Γ_0 is the damping term at zero wave vector (related to the FMR linewidth at low power) and U is an oscillator strength. The first-order Suhl effect is illustrated in figure 7.2(a). It involves the $\mathbf{q}_p = 0$ pumping mode coupling directly to the $\mathbf{q} = 0$ uniform magnon, which then decays through the three-magnon splitting term in $H^{(3)}$ into a pair of magnons with wave vectors $\mathbf{k}$ and $-\mathbf{k}$, each with frequency $\omega_p/2$. From equation (7.12) the part of $H^{(3)}$ responsible for this process is

$$\sum_{\mathbf{k} \neq 0} A^{(3)}(\mathbf{k}, -\mathbf{k}) b^\dagger(\mathbf{k}) b^\dagger(-\mathbf{k}) b(0). \tag{7.19}$$

In the macroscopic description for a bulk ferromagnet it may be concluded (see, e.g., [7, 9]) that the amplitude term $A^{(3)}(\mathbf{k}, -\mathbf{k})$ is approximately proportional to $\omega_M \sin(2\theta) \exp(-i\phi)$. This is largest when the polar angle $\theta = \pi/4$ for the propagating magnons, as depicted on the right in figure 7.2(a).

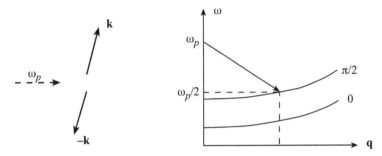

Figure 7.3. The same as in figure 7.2 but for the case of parallel pumping. After Cottam and Tilley [1].

The second-order Suhl effect, by comparison, is shown in figure 7.2(b). Two uniform $\mathbf{q} = \mathbf{0}$ magnons are produced from two pumping modes (each with the frequency ω_p). In this case they decay via the four-magnon scattering process into a magnon pair with wave vectors $\mathbf{k}$ and $-\mathbf{k}$. The last step corresponds to the terms

$$\sum_{\mathbf{k} \neq 0} A^{(4)}(\mathbf{k}, -\mathbf{k}, 0) b^\dagger(\mathbf{k}) b^\dagger(-\mathbf{k}) b(0) b(0) \tag{7.20}$$

as part of equation (7.13). In a macroscopic bulk ferromagnet this effect is greatest when the magnons, which have frequency equal to ω_p, are at the bottom of the band, i.e., with $\theta = 0$, as shown in the figure.

Finally for parallel pumping it is found that the lowest-order term (apart from a constant) in the coupling $H_{\mathrm{p},\parallel}$ is quadratic in the boson operators, by contrast with equation (7.17). The important term has the form

$$H_{\mathrm{p},\parallel} \propto h_{\mathrm{p},\parallel} \sum_{\mathbf{k}} \left[b^\dagger(\mathbf{k}) b^\dagger(-\mathbf{k}) \rho(\mathbf{k}) \exp(-i\omega_\mathrm{p} t) + \text{H.c.} \right]. \tag{7.21}$$

The weighting factor $\rho(\mathbf{k})$ in the macroscopic model is $\omega_\mathrm{M} \sin^2 \theta \exp(-2i\phi)/\omega_\mathrm{B}(\mathbf{k})$, which has its maximum when $\theta = \pi/2$ corresponding to the top of the magnon band. This is illustrated in figure 7.3 and involves a direct excitation of the magnon pair by the pumping mode.

7.1.3 Solitons: introduction and sine-Gordon equation

When nonlinearities are included in the relevant wave equation, some partial differential equations involving position and time as the variables may have types of solutions known as *solitary waves* that have no counterpart in the linear case. In essence, a solitary wave is a disturbance that travels without change of shape. If there is the further property that after two such solitary waves collide (pass through one another) they emerge without change of shape, then they are called *solitons* (see, e.g., Drazin and Johnson [20], and Infeld and Rowlands [21] for general accounts). Often, however, the terms solitary wave and soliton tend to be used interchangeably.

An early observation of a solitary wave was by John Scott Russell in 1834 as a propagating water wave along a canal [21]. The wave was launched from the bow of a canal boat when the boat suddenly stopped. A 'hump' of water detached itself and travelled a large distance along the canal (followed by Scott Russell) without noticeably changing its profile. Later studies of the hydrodynamics of surface waves showed that the phenomenon is governed by a specific NL differential equation, now known as the the Korteweg–de Vries equation, that permits solitary wave solutions. An essential feature, both here and in other examples to be described shortly, is that nonlinearities must be present (playing a role as a 'focusing' effect) to counter-balance the broadening or spreading of the wave that would otherwise take place due to dispersion. The two NL equations capable of soliton solutions that we shall discuss in the context of excitations in solids are the sine-Gordon (sG) equation and the NL Schrödinger equation (NLSE). Both have been studied mathematically in great detail and we will give examples of how they provide an approximate physical description for some real systems.

We start with the sG equation, for which the soliton-related solutions have been applied to domain wall behaviour in magnetic materials and to excitations in certain 1D (chain) ferromagnets. The sG equation for a variable $\phi(z, t)$ as a function of a spatial coordinate z and time t takes the form

$$\frac{\partial^2 \phi}{\partial z^2} - \frac{1}{c^2}\frac{\partial^2 \phi}{\partial t^2} = m^2 \sin \phi. \tag{7.22}$$

Here c is a characteristic velocity and m is a constant that introduces the nonlinearity on the right-hand side. If $m = 0$ we simply have the linear wave equation (e.g., as for electromagnetic waves).

As a physical illustration for the use of the sG equation we mention soliton calculations by Mikeska [22] applied to a 1D (chain) magnet with easy-plane single-ion anisotropy. The model spin Hamiltonian is

$$H = -J \sum_n \mathbf{S}_n \cdot \mathbf{S}_{n+1} + D \sum_n (S_n^z)^2 - g\mu_B B_0 \sum_n S_n^x, \tag{7.23}$$

where $\mathbf{S}_n$ is the spin at position $z = na$ (n = integer), a is the lattice parameter and J is the nearest-neighbour ferromagnetic exchange. For the anisotropy coefficient $D > 0$ the spins prefer an orientation in the xy-plane, while the applied field B_0 stabilizes the alignment along the x-direction. The spin vectors, treated classically, can be written in polar coordinates as $\mathbf{S}_n = S(\sin \theta_n \cos \phi_n, \sin \theta_n \sin \phi_n, \cos \theta_n)$, so the equilibrium state corresponds to $\theta_n = \pi/2$ and $\phi_n = 0$. The linear SWs for small oscillations in the angles about these values can be studied using the methods described in chapter 3.

To show that there are also soliton solutions, in which the angles undergo large variations, Mikeska formed the equations of motion satisfied by the angles and then rewrote them by taking a continuum limit, in which $\phi_n(t)$ is replaced by $\phi(z, t)$, etc. He found that this led to an sG equation for $\phi(z, t)$ in the same form as equation (7.23) with the definitions $c = aS(2DJ)^{1/2}$, $m^2 = \mu_B B_0 / SJa^2$ and $\theta = (1/2SD)(\partial\phi/\partial t)$. The relevant soliton solution of equation (7.23) can be written as

$$\cos[\phi(z, t)] = 1 - 2 \operatorname{sech}^2[m\gamma_0(z - ut - z_0)], \tag{7.24}$$

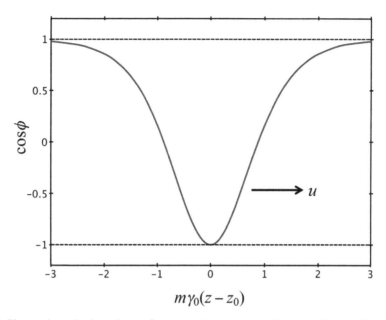

Figure 7.4. Plot to show the dependence of $\cos\phi$ on $(z - z_0)$ at $t = 0$ as given by equation (7.24) for a magnetic-chain soliton as described by the sG equation.

where z_0 denotes the centre position at $t = 0$. The propagation velocity is u ($<c$) and we denote $\gamma_0 = [1 - u^2/c^2]^{-1/2}$. In figure 7.4 a plot is given to illustrate the dependence of $\cos\phi$ on $(z - z_0)$ at $t = 0$. There are large oscillations in ϕ, which switches from being close to zero ($\cos\phi \sim 1$) well away from $z = z_0$ to the opposite direction with $\phi = \pm\pi$ at $z = z_0$. It follows that the polar angle θ also undergoes large oscillations.

The above formalism was applied by Mikeska [22] to $CsNiF_3$, which is fairly well described by the Hamiltonian in equation (7.23). At around the same time neutron-scattering measurements provided good evidence of soliton-like behaviour in $CsNiF_3$ [23]. The sG equation can also be used to analyse magnetic domain-wall dynamics and magnetization switching.

7.1.4 Solitons: the nonlinear Schrödinger equation

The NLSE also has soliton solutions with the applications occurring both in NL optics (e.g., pulses in optical-fibre communications) and in the dynamical properties of magnetic films. A standard (dimensionless) form of writing the NLSE for a variable $u(\xi, \tau)$ is

$$i\frac{\partial u}{\partial \tau} = -\frac{1}{2}\frac{\partial^2 u}{\partial \xi^2} - |u|^2 u. \tag{7.25}$$

This may be compared with the 1D Schrödinger time-dependent equation of quantum mechanics, which may be stated as $i\hbar\partial\psi/\partial t = -(\hbar^2/2\,m)(\partial^2\psi/dx^2)$, where ψ is the wave function, m is the particle mass and the potential function is set to zero. We see

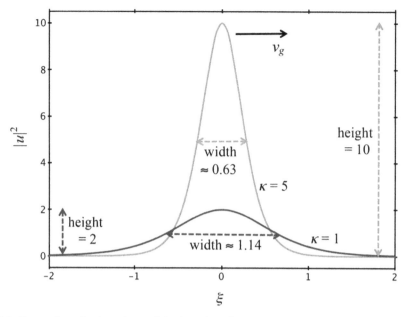

Figure 7.5. Plot to show the dependence of the intensity $|u^2|$ on ξ at $\tau = 0$ as given by equation (7.26) for solitons as described by the NLSE for $\kappa = 1$ and $\kappa = 5$. The behaviour for the pulse height and width is shown for each case.

that the last term of equation (7.25) provides the nonlinearity, while ξ and τ are analogous here to the space-like and time-like variables of Schrödinger's equation.

Solutions of the NLSE are discussed in most textbooks on NL dynamics (see, e.g., [20, 21]). The simplest solution is for one soliton and can be written as

$$u(\xi, \tau) = (2\kappa)^{1/2}\mathrm{sech}[(2\kappa)^{1/2}\tau]\exp(i\kappa\tau), \qquad (7.26)$$

where κ may take any real positive value. The intensity of the pulse, which is proportional to $|u^2|$, has the form of a 'sech-squared' hump when plotted against the spatial variable ξ, as illustrated in figure 7.5 for two cases. In these dimensionless units the pulse height is 2ξ and the pulse width (full width at half-maximum) is $(2/\kappa)^{1/2}$.

Generalizations of the one-soliton solution to equation (7.25) were made by Satsuma and Yajima [24] using a powerful technique known as the inverse scattering method (see, e.g., [20, 21]). The additional solutions differ from that quoted above in that they are 'breathing' modes, meaning that the initial pulse shape changes with distance but re-forms unchanged after a certain distance.

A. NLSE in NL optics

In non-magnetic low-dimensional structures the propagation of solitons generated by a tunable laser in a single-mode quartz fibre was studied experimentally by Mollenauer *et al* [25]. The nonlinearity that provides the connection to the NLSE in this case arises via the Kerr effect discussed in subsections 1.1.2 and 7.1.1. The derivation usually proceeds by starting with Maxwell's equations with a NL polarization term and then forming a wave packet which is expanded using the *slowly varying amplitude* (SVA) approximation. The main steps are summarized below, but a full account can be found in [1].

Maxwell's equations for an isotropic NL material can be expressed as

$$\frac{\partial^2 E}{\partial z^2} - \mu_0 \frac{\partial^2 D}{\partial t^2} = \mu_0 \frac{\partial^2 P^{NL}}{\partial t^2}, \tag{7.27}$$

where the spatial variations are taken to be in just one direction z. The D field here contains only the linear polarization and so $D = \varepsilon_0 \varepsilon(\omega) E$ for the Fourier components at frequency ω and P^{NL} is the NL polarization. For applications to a wave packet we write

$$E(z, t) = F(z, t)\exp(iq_0 z - i\omega_0 t), \tag{7.28}$$

with q_0 and ω_0 denoting the central wavenumber and frequency of the packet. The above relationships enable the derivative terms on the left-hand side of equation (7.27) to be formally rewritten in terms of the envelope function F and its derivatives. In particular, we find

$$-\mu_0 \frac{\partial^2 D}{\partial t^2} = \exp(iq_0 z - i\omega_0 t)\int_{-\infty}^{\infty} (q_0 + k)^2 F(z, \nu)\exp(-i\nu t)\mathrm{d}\nu, \tag{7.29}$$

where $q_0 + k$ means the wavenumber corresponding to frequency $\omega_0 + \nu$. The SVA approximation mentioned earlier is basically just the assumption that $|k| \ll q_0$ for the important terms in the integral, so a Taylor series expansion gives $k \simeq k'\nu + (1/2)k''\nu^2$, denoting

$$k' = \frac{\mathrm{d}k}{\mathrm{d}\omega} = \frac{1}{v_\mathrm{g}}, \qquad k'' = \frac{\mathrm{d}^2 k}{\mathrm{d}\omega^2} = \frac{\mathrm{d}}{\mathrm{d}\omega}\left(\frac{1}{v_\mathrm{g}}\right). \tag{7.30}$$

Here the derivatives are evaluated at $\omega = \omega_0$ and v_g is the group velocity of the wave packet. With the above approximations and similarly assuming that the frequency spread around ω_0 is small, we find after some algebra that the envelope function F satisfies

$$i\left(\frac{\partial}{\partial z} + \frac{1}{v_\mathrm{g}}\frac{\partial}{\partial t}\right)F = \frac{k''}{2}\frac{\partial^2 F}{\partial t^2} - k_0\rho \, |F|^2 F. \tag{7.31}$$

where we have assumed a Kerr nonlinearity to replace P^{NL} and ρ denotes the third-order NL coefficient defined in equation (1.18). The terms on the left-hand side describe an envelope travelling at speed v_g, so it is convenient to make a transformation to new variables by $(z, t) \rightarrow (Z, T)$ where $Z = z - v_\mathrm{g}t$ and $T = t$, giving

$$i\frac{\partial F}{\partial Z} = \frac{k''}{2}\frac{\partial^2 F}{\partial T^2} - k_0\rho \, |F|^2 F. \tag{7.32}$$

We may now compare the above result with the previous canonical form of the NLSE quoted in equation (7.25). Apart from the scaling of the coefficients, we see that the equations are essentially the same with two provisions: first, the roles of dimensionless ξ and τ in (7.25) are replaced by T and Z, respectively, and second, the coefficients k'' and ρ must have opposite signs. The first of these points (the interchange of space-like and time-like variables) is of no particular significance, but the second point gives us as a necessary condition for solitons the so-called *Lighthill criterion* that

$$k''\rho < 0. \tag{7.33}$$

This ensures the physical property mentioned earlier that the spreading effect of a soliton pulse due to dispersion must be counterbalanced by the nonlinearity.

B. NLSE for magnetic films

Another important application of the NLSE involves magnetic systems, notably the propagation of envelope solitons in ferromagnetic thin films where experiments were first reported by Kalinikos *et al* [26] for samples of YIG. As in the NL optics case we present just an outline of the main steps giving the connection with the NLSE, while fuller descriptions can be found in [1, 15].

Considering, for simplicity, the situation where the nonlinearities arise only from the four-magnon interaction discussed earlier, we find the operator equation of motion for $b_\ell(\mathbf{q}_\|)$ is

$$\left(\frac{\partial}{\partial t} + \omega_\ell\left(\mathbf{q}_\|\right)\right)b_\ell\left(\mathbf{q}_\|\right) + i \sum_{\ell_1,\ell_2,\ell_3} \sum_{\mathbf{q}_{\|1},\mathbf{q}_{\|2},\mathbf{q}_{\|3}} A^{(4)}_{\ell_1,\ell_2,\ell_3}\left(\mathbf{q}_{\|1}, \mathbf{q}_{\|2}, \mathbf{q}_{\|3}\right)$$

$$\times b^\dagger_{\ell_1}\left(\mathbf{q}_{\|1}\right)b_{\ell_2}\left(\mathbf{q}_{\|2}\right)b_{\ell_3}\left(\mathbf{q}_{\|3}\right)\delta_{\mathbf{q}_\|+\mathbf{q}_{\|1},\mathbf{q}_{\|2}+\mathbf{q}_{\|3}} = 0. \tag{7.34}$$

This is obtained using the quantum-mechanical operator equation (1.36) and we follow the notation for dipole-exchange films in subsection 3.3.2, so there is a band label ℓ and and an in-plane 2D wave vector $\mathbf{q}_\|$. The expression used for the $H^{(4)}$ interaction is analogous to equation (7.13). We now study the evolution of a narrow wave packet of magnons with central in-plane wave vector $\mathbf{q}_{\|0}$ and central frequency $\omega_{\ell0}$. The conservation laws for the four-magnon processes within this packet yield $2\omega_{\ell0} = \omega_{\ell1} + \omega_{\ell2}$ for the same SW branch and $2\mathbf{q}_{\|0} = \mathbf{q}_{\|1} + \mathbf{q}_{\|2}$. Approximate forms of the equations of motion of each of the related operators $b_\ell(\mathbf{q}_{\|0})$ and $b_\ell(\mathbf{q}_{\|i})$ (with $i = 1, 2$) can then be obtained by retaining only the important terms in equation (7.34).

A conclusion is that the four-magnon interaction leads to a modification (or 'renormalization', in the language of many-body theory) of the frequencies for the magnons in the above process. Specifically it is found (see, e.g., [1, 15]) that $\omega_{\ell0} \to \Omega_{\ell0}$ and $\omega_{\ell i} \to \Omega_{\ell i}$, where

$$\Omega_{\ell0} = \omega_{\ell0} + A_{00}\,|b_{\ell0}|^2, \qquad \Omega_{\ell i} = \omega_{\ell i} + 2A_{i0}\,|b_{\ell0}|^2 \quad (i = 1, 2). \tag{7.35}$$

Here A_{00} and A_{i0} are shorthand for particular matrix elements of the interaction $A^{(4)}$; their definitions are of little concern to us here. What is important is that the NL dispersion relations $\Omega_{\ell i}$ can be expanded in a Taylor series about the central wave vector $\mathbf{q}_{\|0}$ of the packet. Thus, putting $\mathbf{q}_{\|i} = \mathbf{q}_{\|0} \pm \mathbf{k}_\|$ and assuming $|\mathbf{k}_\||$ is small, we may deduce that

$$\Omega_{\ell1,2} = \omega_{\ell1,2} \pm v_g k_\| + (1/2)Dk_\|^2 + \cdots + 2A_{00}\,|b_{\ell0}|^2, \tag{7.36}$$

where the group velocity v_g has been introduced in the same manner as for the preceding NL optics case (see equation (7.30)). Also $D = \partial^2\omega_\ell(\mathbf{q}_\|)/\partial q_\|^2$, evaluated at $k_\| = 0$, is the dispersion term in the same notation as for chapter 3.

Equations (7.35) and (7.36), taken together with the four-magnon conservation condition, gives as a requirement for the parametric instability to occur:

$$Dk_\|^2\left(Dk_\|^2 + 4A_{00}\,|b_{\ell0}|^2\right) < 0. \tag{7.37}$$

By inspection, this can only be true if D and A_{00} are opposite in sign, i.e., we have $A_{00}\,D < 0$ which is just another form of the Lighthill criterion given in equation (7.33). Then, by following the analogous steps for the wave packet as in the NL case earlier, we arrive at an NLSE in the form

$$\left[i\left(\frac{\partial}{\partial t} + \frac{1}{v_g}\frac{\partial}{\partial \zeta}\right) + \frac{D}{2}\frac{\partial^2}{\partial \zeta^2} - A|\psi_\ell|^2\right]\psi_\ell(\zeta, t) = 0, \tag{7.38}$$

where ζ is an in-plane spatial variable and we define $A = 2M_0 A_{00}/\gamma$. The transverse magnetization term ψ_ℓ, which plays an analogous role to the envelope function F in equation (7.31), is formed by writing

$$\psi_\ell(\zeta, t) = \left(\frac{\gamma}{2M_0}\right)^{1/2}\int dq\ b_\ell\left(q + q_{\|0}, t\right)\exp[iq\zeta - i\omega_{\ell0}t], \tag{7.39}$$

Although we have been able to demonstrate here the formal, qualitative similarities between the envelope solitons in optics and magnetic films, there are major differences at a quantitative level. For example, the pulse width dispersions in optics and magnetics (taking YIG as a benchmark) are 1 ps and 100 ns, respectively, and the group velocities are of order 2×10^8 ms^{-1} and 2×10^4 ms^{-1}, respectively.

7.2 Applications to non-magnetic low-dimensional systems

In this section we will focus on three main areas of application: NL optics via surface modes, Kerr effect bistability in layered and multilayered systems, and gap solitons in NL optics. These applications involve the third- and fourth-order susceptibility tensors described in subsection 7.1.1.

7.2.1 Nonlinear optics via surface modes

The NL effects mentioned earlier can lead to enhancement of the electric field (due to the higher powers of E) when measuring transmission through a suitable sample, especially when multiple interfaces are involved. The most basic implementation of these ideas is for transmission through a Fabry–Pérot resonator, in which a film of thickness L is coated with partially reflecting mirrors. The use of a Fabry–Pérot in the *linear* regime is familiar and widespread (see the general optics textbook by Born and Wolf [27]). It is, for example, the essential tool in BLS in solids, as mentioned in section 1.3. The calculation of the transmission spectra in a normal-incidence geometry is standard in linear optics, giving repeated peaks when the condition $q_B L = n\pi$ is satisfied, where n is any positive integer and q_B is the optical wave-number in the film. The generalization to include the simplest NL effects of SHG through terms such as $\chi_{ijk}^{(2)}(-2\omega; \omega, \omega)$ is surprisingly complicated as a result of all the internal reflections and transmissions in this geometry. A fairly complete and detailed account is presented by Hashizume *et al* [28].

Another structure that has been studied for the role of SHG on the reflectance and transmission is the periodic superlattice, where one or more of the components is NL. Notably, calculations have been made by Agranovich and Kravstov [29], and Boyd and Sipe [30]. They were able to take advantage in some cases of a generalized form for the effective-medium approximation, which was described in chapter 5 for the linear regime. Experiments due to Fischer *et al* [31] have provided quantitative confirmation for many of the predictions.

A more interesting, and surface specific, illustration of field enhancement is through NL coupling to surface modes such as surface plasmons or surface polaritons in bounded media. For significant NL effects to be observable it is important that the surface mode be deeply penetrating and for this reason the LRSP that was mentioned briefly in chapter 5 is suitable. This mode was originally predicted by Sarid [32] to occur for a thin metallic layer (electron gas) with a linear dielectric function bounded on each side by a non-lossy medium. In later work the theory was extended to NL optics [34]. Experiments to demonstrate NL optical effects using the LRSP were reported by Quail *et al* [33]. An example of their measurements of the SHG reflection coefficient is shown in figure 7.6. The active medium was an Ag film and the experimental set-up is given in the inset. Theory and experiment are seen to be in good agreement.

7.2.2 Kerr effect bistability in multilayers

We now discuss consequences of the Kerr effect in multilayer geometries, starting with the example of a NL surface excitation given near the beginning of this book in section 1.1.2. There we outlined a calculation for a NL material having a dielectric function with the form of equation (1.18) filling the half-space $z < 0$, while the other half-space ($z > 0$) corresponds to a linear material. Following analogies with self-focusing, we showed that a NL wave is obtained in s-polarization for the electric field localized near the $z = 0$ interface. The form of the solution is given explicitly by equations (1.26) and (1.27).

A few further steps are required in order to arrive at a dispersion relation. First, the usual electromagnetic boundary conditions on the field components can be applied at $z = 0$. These are equivalent to requiring that E and dE/dz are continuous at the boundary and these lead to the condition

$$\kappa \tanh(\kappa z_0) = \kappa_1, \tag{7.40}$$

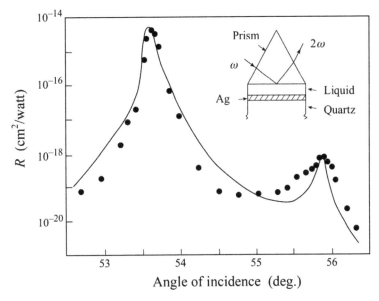

Figure 7.6. Second-harmonic reflection coefficient R for SHG via the LRSP in an Ag film plotted versus the angle of incidence. The solid lines are theory and the points represent experimental data. The experimental geometry is shown in the inset. After Quail *et al* [33].

where κ and κ_1 were introduced earlier and z_0 is the remaining constant of integration. Equation (7.40) implies $\kappa_1 < \kappa$ (or $\varepsilon_1 > \varepsilon$) as a necessary inequality. The standard procedure (as discussed fully in [1], for example) is to utilize the time-averaged power flow P as a basic parameter. The power flow is obtained from the Poynting vector $\mathbf{E} \times \mathbf{H}$, which may be evaluated in terms of z_0 and other quantities. Eventually the dispersion relation is found from equation (7.40) in the form

$$P = P_0\left[1 + \frac{1}{2}\left(\frac{\kappa}{\kappa_1} + \frac{\kappa_1}{\kappa}\right)\right] > 2P_0, \qquad (7.41)$$

where $P_0 = c^2 q_x \kappa / \mu_0 \rho \omega^3$. This result clearly establishes that there is a critical power level for the NL mode to exist. Further discussion and some minor extensions are provided in [35, 36].

Subsequently the analytic theory of s-polarized NL polaritons in general multi-layers (including superlattices) was developed by Vassiliev and Cottam [37], including the situation where all layers may have a Kerr-type dielectric function. Their results are illustrated for a special case of a NL film bounded on each side by the same NL cladding material in figure 7.7. The results are plotted in terms of an effective refractive index $n = cq_x/\omega$ versus $p^{1/2}$, where $p = \rho E_1^2/2$ is a scaled nonlinearity coefficient. Here E_1 denotes the magnitude of the electric field at either of the interfaces with the film; it is the additional parameter to be related to the power flow (just as we had z_0 in the previous calculation). The film thickness has been chosen as one half of the vacuum wavelength λ_0 of the light. The modes labelled 'Z' are branches to the dispersion that occur even in the limit of a linear cladding material,

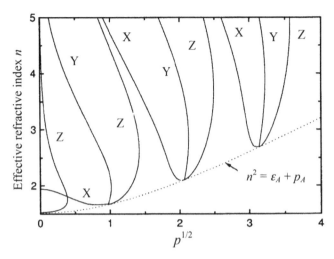

Figure 7.7. Dispersion curves for s-polarized polaritons in the case of an NL film bounded symmetrically by two NL cladding materials. The plots are shown as an effective refractive n for the film versus $p^{1/2}$, taking $\varepsilon_A = 1$ for the bounding material and $p/p_A = 2$ for the ratio of NL coefficients. Other labels and notation are explained in the text. After Vassiliev and Cottam [37].

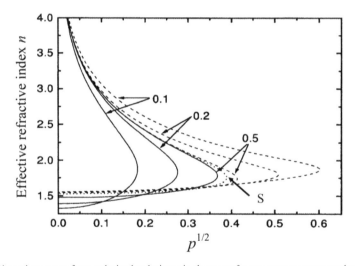

Figure 7.8. Dispersion curves for s-polarized polaritons in the case of a two-component superlattice when both constituents A and B have NL dielectric functions. The numerical values correspond to d_A/λ_0 and other parameters are given in the text. Curves for the Bloch factor equal to -1 and 1 are shown by solid and dashed lines, respectively. The plots are plotted as an effective refractive n versus $p^{1/2}$. Curve 'S' shows a corresponding single-layer branch for comparison. After Vassiliev and Cottam [37].

whereas 'X' and 'Y' refer to the additional branches when the cladding material is NL. These results exhibit the property of multistability due to the NL effects, i.e., n is a multivalued function of p for certain ranges of the parameters.

A further example is given in figure 7.8, which is for a two-component periodic superlattice, where both constituents A and B are NL. Although there is no rigorous analogue of Bloch's theorem in the NL regime, we can nevertheless examine the behaviour for particular values of the Bloch factor $\exp(iQD)$, following our previous notation, while other types of modes are not necessarily precluded. The assumed parameters are $\varepsilon_A = 2.3$, $\varepsilon_B = 2.45$, $\rho_A/\rho_B = 0.5$ and $d_B/\lambda_0 = 0.5$. Here the nonlinearity parameter p is defined as $\rho_B E_1^2/2$, where E_1 is the electric-field magnitude at an A–B interface.

In recent years there have been continuing studies of various types of polaritons in layered structures, e.g., taking characteristic frequency dependences of the linear and NL dielectric functions into account and studying other electric-field polarizations (see [38] and references therein).

7.2.3 Gap solitons in nonlinear optics

We described in subsection 7.1.4 how the Lighthill criterion $k''\rho < 0$) provides a necessary condition for formation of solitons in terms of the NLSE. Usually the nonlinearity coefficient ρ is positive, so typically it is the curvature term in the dispersion, $k'' = d^2k/d\omega^2$ that must be negative for optical solitons. One way in which this can be engineered relates to wave propagation in periodic structures such as superlattices. For example, we have seen in several of the previous chapters that band gaps (or stop bands) open up at the mini-Brillouin-zone boundaries and that the curvature in the dispersion has opposite signs just below and just above the gaps, where it changes rapidly. We may see this property, for example, in figure 1.9 for electromagnetic waves, figure 2.18 for phonons and figure 3.23 for magnons. It follows, therefore, that the conditions for soliton propagation might be realized near to these band gaps, leading to the concept of *gap solitons*.

The conditions for gap solitons in NL optics were first investigated by Mills and co-workers [39, 40]. They assumed a periodic two-component superlattice with a large (but finite) number of layers. One constituent material was taken to be linear while the other was assumed to have a Kerr-type NL dielectric function with the same form as in previous sections. The formation of a soliton wave packet was confirmed and the envelope function was found to be close to the sech-squared profile as plotted in figure 7.5. Theoretical studies were also carried out to show the possibility of gap solitons in more complicated layered structures, such as those based on Fibonacci growth rules (see figure 2.13(a)). In particular, Dutta Gupta and Ray [41] calculated the transmission through 55-layer and 233-layer sequences taking both the A and B building blocks to have a Kerr-type nonlinearity. They found results that were qualitatively rather similar to those in [39, 40].

Shortly after the calculation applied to the periodic superlattices the first experimental evidence was published by Mohideen *et al* [42]. Their measurements were made on a grating that was produced by ultraviolet laser interference on a GeO$_2$-doped SiO$_2$ glass fibre. By varying the longitudinal strain in the fibre they were able to adjust the grating period and hence achieve tunability. Further discussion of these results can be found in [1]. The up-to-date review article on gap solitons in NL optics by Maimistov [43] covers the recent work in this field, including developments that have been mainly on the experimental side and with regard to the materials technology.

7.3 Applications to magnetic low-dimensional systems

As in the previous section on non-magnetic systems we will highlight a few selected topics here for applications to NL magnetic low-dimensional systems. They all involve the formalism described in section 7.1 and comprise: decay and instability processes for magnons, resonant frequency multiplication for magnons and the Bose–Einstein condensation (BEC) in a magnon gas. The emphasis will be on low-dimensional structures such as thin films, nanowires and small spheres.

7.3.1 Decay and instability of magnons

The three- and four-magnon processes discussed earlier can give rise to a variety of physical effects and we start here with some results for the damping (or reciprocal lifetime) of the magnons. In principle, this may be measured as the non-instrumental part of the line width in FMR or inelastic light scattering experiments. The behaviour in bulk materials is very well understood (see, for example, the review article by Keffer [7]), but in low-dimensional structures the magnon-interaction amplitudes and the overall conservation conditions are modified.

The basic processes, however, are the same. Thus the simplest contribution to the damping comes from the three-magnon splitting and confluence depicted in figures 7.1(a) and (b). Specifically a magnon can split into two other magnons, which can subsequently recombine, representing a decay channel for the initial magnon provided conservation of energy and (sometimes) wave vector are satisfied. In bulk systems the 3D wave vector must always be conserved, but for films or wires only those wave-vector components along the directions of translational symmetry are conserved. The damping is of second order in the three-magnon interaction, i.e., proportional to $|A^{(3)}|^2$ in our previous notation. The four-magnon process can similarly mediate a damping process in which there are three intermediate (or 'virtual') magnons and again it is subject to conservation of energy and conservation of wave vector in the directions of translational symmetry. Typically the simpler three-magnon processes will provide the dominant contribution *unless* the magnetic dipole–dipole interactions are negligibly small (as in the Heisenberg limit) or when the conditions for wave-vector conservation cannot be satisfied.

Magnetic nanowires form an ideal system for studying the damping, since the magnon spectrum consists of discrete branches due to the lateral spatial quantization (see section 3.5) and the wave vectors are conserved only in 1D, i.e., along the length of the wires. Measurements of the damping were reported by Boone *et al* [44] for nanowires (or stripes) with a rectangular cross section. They used a spin-torque FMR technique with permalloy nanowire stripes of thickness 6 nm and different widths in the range 100–250 nm. They obtained the linear magnon frequencies as a function of applied magnetic field (along the length of the nanowires), which consisted of discrete (or quantized) branches, and they deduced the behaviour of the damping for the lowest set of magnons. Some results for the field dependence of the damping of the lowest-frequency magnon are shown in figure 7.9 for nanowires with widths of 140 and 125 nm. A strong field dependence was observed, particularly for the wider wire, and there were peaks (indicated by the arrows) at an intermediate

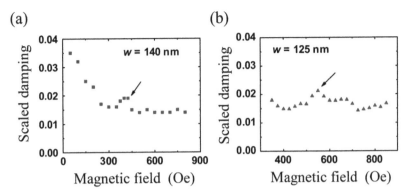

Figure 7.9. Scaled damping parameter for the lowest-frequency magnon modes in permalloy nanowires with width w equal to (a) 140 nm and (b) 125 nm, showing the dependence on applied magnetic field (100 Oe is equivalent to 0.01 T). Arrows indicate intermediate peaks in the damping. After Boone *et al* [44].

field. The damping is presented in terms of scaled dimensionless units after making allowances for inhomogeneous broadening in the experiments [44]. The observed damping contributions were ascribed as being due mainly to three-magnon processes in conjunction with effects due to roughness at the lateral edges of the wires. The roughness degrades slightly the property of translational symmetry along the length of the wires and so magnons can decay via direct scattering into another magnon branch with the same energy (but different wave vector). A theory for three-magnon and four-magnon damping in rectangular nanowires was subsequently developed by Nguyen *et al* [45]. It did not include edge roughness effects but could account, at least qualitatively, for the observed field-dependent effects arising from the three-magnon processes.

Next we turn to some examples of SW instabilities under conditions of microwave pumping at elevated power levels. The background theory has already been covered in subsection 7.1.2, mainly in the context of thin films with in-plane direction of magnetization. It can, however, be extended to other sample geometries where there may be a component of the static magnetization perpendicular to the surface, e.g., as in spheres, flat disks and wires. In figure 7.10 some plots of the threshold microwave-field amplitude h for instability versus applied magnetic field $H_0 = (B_0/\mu_0)$ are shown, using data from Rezende *et al* [46] for a YIG sphere. Results for parallel pumping and perpendicular pumping for the first-order Suhl effect are in panels (a) and (b), respectively. In each case the lower curve shows the instability threshold, while the upper curve corresponds to the onset of *auto-oscillations*, which are a further stage of instability acting as a precursor to chaotic behaviour. The instability threshold in figure 7.10(a) for parallel pumping follows the characteristic *butterfly curve*. For applied fields below a critical value at H_c the excitation of parametric SWs at frequency $\omega/2$ is energetically possible, such as the depiction in figure 7.3 for a film, but this is no longer possible when $H_0 > H_c$ so the instability threshold rises sharply. For perpendicular pumping there is also a critical field as indicated, but the shape of the curve is quite different in the case of a sphere rather than a film [46].

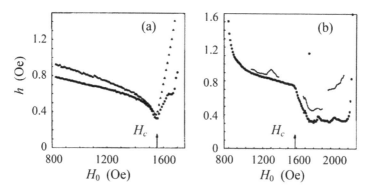

Figure 7.10. Threshold values of the microwave pumping-field amplitude h plotted against the static applied field $H_0 = B_0/\mu_0$ in a 1 mm diameter YIG sphere for (a) parallel pumping at 9.5 GHz with H_0 along [111] and (b) first-order Suhl effect at 8.87 GHz with H_0 along [110]. The lower curves refer to the initial instability thresholds and the upper curves (or fragments) refer to auto-oscillations. After Rezende *et al* [46].

For smaller-size samples, with dimensions of the order of nanometres or micrometres rather than millimetres, the spatial quantization of the magnon modes becomes more pronounced and consequently the shape of the butterfly curve is expected to be modified, showing sharp structural features (see, e.g., [47, 48]). Another interesting development involves experiments by Sandweg *et al* [49] to detect by spin pumping the parametrically excited exchange magnons, rather than the dipole-exchange modes as described earlier. Specifically the magnons were injected into a YIG film by parametric pumping with a microwave field and the inverse spin-Hall effect was used to detect a voltage. Magnon wavelengths down to the sub-micron level were achievable. This opens new perspectives for applications in spintronics (see subsection 3.7.2 for a general discussion of this topic).

7.3.2 Frequency multiplication for magnons

The three- and four-magnon NL processes described earlier make frequency multiplication for the magnons allowable. For example, if two magnons of frequency ω_0 participate in the confluence process sketched in figure 7.1(b), corresponding to an interaction amplitude proportional to $A^{(3)*}$ in equation (7.12), a magnon of frequency $2\omega_0$ will be produced. Similarly, the four-magnon interactions can lead to output of magnons with frequency $3\omega_0$ through the amplitude term $B^{(4)*}$ in equation (7.13). It then becomes a matter of how to achieve this in practical terms and how to maximize the effect.

Recently Demidov *et al* [50] demonstrated experimentally the effect of resonant frequency multiplication using sub-micron sized permalloy dots which were subjected to an intense microwave magnetic field. They observed harmonics corresponding to output at two and three times the frequency of the applied signal frequency. The magnetic dots were produced starting from a 10 nm thick permalloy film, which was then patterned by electron-beam lithography to form the elliptical dots with semi-major and semi-minor axes of length 500 and 250 nm, respectively. It was found that the efficiency of signal multiplication is strongly enhanced when the harmonic is arranged to be resonant with the (linear) normal dynamical modes of the dot, which were studied separately using BLS at lower power levels. This condition of achieving resonance with the linear modes is different for the double- and triple-frequency harmonics. As a consequence, Demidov *et al* found that their elliptical sample shape was particularly efficient for resonant frequency tripling. Their results may be an important step for the implementation of nanoscale frequency multipliers for integrated microwave technology.

7.3.3 Bose–Einstein condensation in a magnon gas

The ideas associated with the BEC were developed by Bose and Einstein separately in the 1920s. In broad terms, a BEC is now taken to refer to a phase or state of matter in which a macroscopically large number of the bosons occupy, or have condensed to, their lowest quantum state (see, e.g., the books by Landau and Lifshitz [51] and Pitaevskii and Stringari [52] for general discussion). The occurrence of BEC requires the (boson) particle density to exceed a critical value, which may be achieved by lowering the temperature or by raising the boson density or by some combination of these effects. Initially superfluids and superconductors provided the only examples of BEC, but relatively recently it was achieved for certain diluted atomic gases at ultralow temperatures [53, 54]. This discovery served both to suggest that BEC might be a more widespread phenomenon than previously believed and to spur on efforts to observe BEC in other physical systems.

A magnon gas in a ferromagnet or ferrimagnet at a temperature well below its Curie temperature T_C presents a particularly promising candidate for BEC formation, since we have already seen that the magnons behave like weakly interacting bosons and their densities can be manipulated and controlled through microwave pumping techniques. Indeed, the observation of BEC in a magnon gas at room temperature was first reported by Demokritov *et al* [55] in 2006 and consolidated in further work by the same group shortly afterwards (see, e.g., [56–58]).

The experiments were carried out using thin films of YIG that were magnetized in-plane by a static applied field and had thicknesses in the 2–10 µm range. The form of the (linear) magnon dispersion relation in the dipole-exchange regime follows that discussed in section 3.3. We note in particular that the lowest-frequency magnons are those excited with in-plane wave vectors $\mathbf{q}_\parallel$ parallel to the magnetization direction and that these magnons have their minimum frequency at some value with $q_\parallel \neq 0$. This feature occurs due to competition at $q_\parallel \neq 0$ between the dipole–dipole interactions which decrease the magnon frequency (as in the case of magnetostatic backward bulk modes in subsection 3.3.1) and the exchange interactions which cause an increase in frequency. In the BEC experiments the frequency minimum occurred at wave vector $q_{\parallel,m} \simeq 2 \times 10^5$ cm^{-1} as verified by the use of BLS spectroscopy.

The magnons are driven by a microwave pumping field at angular frequency ω_p until, when a threshold high power level is exceeded, magnon pairs at frequency $\omega_p/2$ and with equal and opposite wave vectors are produced parametrically as explained before. This stage is followed by a rapid redistribution of magnon energies (through scattering events) going down to the frequency minimum $\omega(q_{\parallel,m})$ for the bosons.

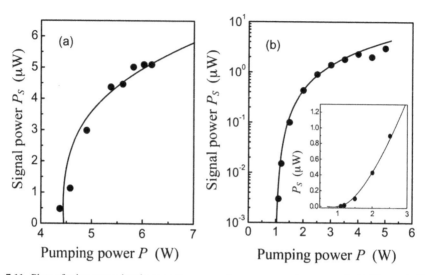

Figure 7.11. Plots of microwave signal power P_S versus microwave pumping power P. (a) Theory (solid line) compared with experimental data (dots) from [57]. (b) The same but using experimental data (see text) from [58]. Also, the inset in (b) shows the behaviour near the threshold value in greater detail. After Rezende [59].

This redistribution occurs on a time scale that is much shorter than the relatively long spin-lattice relaxation time, so the 'hot' magnon gas state persists. In contrast, if the pumping signal power exceeds another threshold level, typically much larger than the one discussed earlier for the SW instability under parallel pumping, the already elevated population of magnons begins to condense in a state around the minimum frequency and to display quantum coherence.

A theory that was successful in explaining many of the observed features of the BEC was developed by Rezende [59]. His model incorporates the cooperative properties of magnons with frequencies close to the dispersion minimum at $\omega(q_{\|,m})$ through the four-magnon interaction term discussed in subsection 7.1.2. The theory accounts well for the onset of the BEC as the microwave power P is increased above a threshold value. Also it predicts coherence for the magnons in the condensate and separation of the magnon gas above threshold into two parts, one in thermal equilibrium and one with coherent magnons. Comparisons between theory and two sets of experimental data (the first obtained with coherent pumping and the second with incoherent pumping) are illustrated in figure 7.11, both yielding good agreement. The signal power relates to the mean power radiated by the uniform magnetization precessing about the static field.

References

[1] Cottam M G and Tilley D R 2005 *Introduction to Surface and Superlattice Excitations* 2nd edn (Bristol: IOP)

[2] Yariv A 1989 *Quantum Electronics* 3rd edn (New York: Wiley)

[3] Butcher P N and Cotter D 1990 *The Elements of Nonlinear Optics* (Cambridge: Cambridge University Press)

[4] Mills D L 1998 *Nonlinear Optics* 2nd edn (Berlin: Springer)

[5] Nye J F 1985 *Physical Properties of Crystals* (Oxford: Clarendon)

[6] Wherrett B S 1986 *Group Theory for Atoms, Molecules and Solids* (Englewood Cliffs, NJ: Prentice-Hall)

[7] Keffer F 1966 *Handbuch der Physik* vol 18 (Berlin: Springer) p 1

[8] L'vov V S 1994 *Turbulence under Parametric Excitation: Applications to Magnets* (Heidelberg: Springer)

[9] Gurevich A G and Melkov G A 1996 *Magnetization Oscillations and Waves* (Boca Raton, FL: CRC)

[10] Stancil D D and Prabhakar A 2009 *Spin Waves: Theory and Application* (Heidelberg: Springer)

[11] Wigen P (ed) 1994 *Nonlinear Phenomena and Chaos in Magnetic Materials* (Singapore: World Scientific)

[12] Cottam M G (ed) 1994 *Linear and Nonlinear Spin Waves in Magnetic Films and Superlattices* (Singapore: World Scientific)

[13] Hillebrands B and Ounadjela K (ed) 2001 *Spin Dynamics in Confined Magnetic Structures* (Berlin: Springer)

[14] Demokritov S O and Slavin A N (ed) 2013 *Magnonics: From Fundamentals to Applications* (Berlin: Springer)

[15] Wu M and Hoffmann A (ed) 2013 *Solid State Physics: Recent Advances in Magnetic Insulators—From Spintronics to Microwave Applications* vol 64 (Amsterdam: Academic)

[16] Bloembergen N and Damon R 1952 *Phys. Rev.* **85** 699

[17] Bloembergen N and Wang S 1954 *Phys. Rev.* **93** 72

[18] Suhl H 1957 *J. Phys. Chem. Solids* **1** 209

[19] Schlömann E, Green J and Milano V 1960 *J. Appl. Phys.* **31** 3865

[20] Drazin P G and Johnson R S 1989 *Solitons: An Introduction* (Cambridge: Cambridge University Press)

[21] Infeld E and Rowlands G 1990 *Nonlinear Waves, Solitons, and Chaos* (Cambridge: Cambridge University Press)

[22] Mikeska H J 1978 *J. Phys. C: Solid State Phys.* **11** L29

[23] Kjems J K and Steiner M 1978 *Phys. Rev. Lett.* **41** 1137

[24] Satsuma J and Yajima N 1974 *Suppl. Prog. Theor. Phys.* **55** 284

[25] Mollenauer L F, Stolen R H and Gordon J P 1980 *Phys. Rev. Lett.* **45** 1095

[26] Kalinikos B A, Kovshikov N G and Slavin A N 1983 *Sov. Phys.—JETP Lett.* **38** 413

[27] Born M and Wolf E 1966 *Principles of Optics* (Oxford: Pergamon)

[28] Hashizume N, Ohashi M, Kondo T and Ito R 1995 *J. Opt. Soc. Am.* B **12** 1894

[29] Agranovich V M and Kravstov V E 1985 *Solid State. Commun.* **55** 85

[30] Boyd R W and Sipe J E 1994 *J. Opt. Soc. Am.* B **11** 297

[31] Fischer G L, Boyd R W, Gehr R J, Jenekhe S A, Osaheni J A, Sipe J E and Weller-Brophy L A 1995 *Phys. Rev. Lett.* **74** 1871

[32] Sarid D 1981 *Phys. Rev. Lett.* **47** 1927

[33] Quail J C, Rako J G, Simon H J and Deck R 1983 *Phys. Rev. Lett.* **50** 1987

[34] Deck R T and Sarid D 1982 *J. Opt. Soc. Am.* **72** 1613

[35] Stegeman G I, Seaton C T, Hetherington W M, Boardman A D and Egan P 1986 *Electromagnetic Surface Excitations* ed R F Wallis and G I Stegeman (Berlin: Springer) p 261

[36] Wendler L 1986 *Phys. Status Solidi* b **135** 759

[37] Vassiliev O N and Cottam M G 2000 *Surf. Rev. Lett.* **7** 89
[38] Dzedolik I V and Karakchieva O 2013 *Appl. Opt.* **52** 3073
[39] Chen W and Mills D L 1987 *Phys. Rev. Lett.* **58** 160
[40] Mills D L and Trullinger S E 1987 *Phys. Rev. B* **36** 947
[41] Dutta Gupta S and Ray D S 1990 *Phys. Rev. B* **41** 8047
[42] Mohideen U, Slusher R E, Mizrahi V, Erdogan T, Kuwata-Gonokami M, Lemaire P J, Sipe J E, de Sterke C M and Broderick N G R 1995 *Opt. Lett.* **20** 1674
[43] Maimistov A 2010 *Quantum Electron.* **40** 756
[44] Boone C T, Katine J A, Childress J R, Tiberkevich V, Slavin A, Zhu J, Cheng X and Krivorotov I N 2009 *Phys. Rev. Lett.* **103** 167601
[45] Nguyen H T, Akbari-Sharbaf A and Cottam M G 2011 *Phys. Rev. B* **83** 214423
[46] Rezende S M, Azevedo A and de Aguiar F M 1994 *Linear and Nonlinear Spin Waves in Magnetic Films and Superlattices* ed M G Cottam (Singapore: World Scientific) p 335
[47] Chartoryzhskii D N, Kalinikos B A and Vendik O G 1976 *Solis State Commun.* **20** 985
[48] Nguyen H T and Cottam M G 2014 *Phys. Rev. B* **89** 144424
[49] Sandweg C W, Kajiwara Y, Chumak A V, Serga A A, Vasyuchka V I, Jungfleisch M B, Saitoh E and Hillebrands B 2011 *Phys. Rev. Lett.* **106** 216601
[50] Demidov V E, Ulrichs H, Urazhdin S, Demokritov S O, Bessonov V, Gieniusz R and Maziewski A 2011 *Appl. Phys. Lett.* **99** 012505
[51] Landau L D and Lifshitz E M 1980 *Statistical Physics* (Oxford: Pergamon)
[52] Pitaevskii L and Stringari S 2003 *Bose–Einstein Condensation* (Oxford: Clarendon)
[53] Anderson M H, Ensher J R, Matthews M R, Wieman C E and Cornell E A 1995 *Science* **269** 198
[54] Davis K B, Mewes M-O, Andrews M R, van Druten N J, Durfee D S, Kurn D M and Ketterle W 1995 *Phys. Rev. Lett.* **75** 3969
[55] Demokritov S O, Demidov V E, Dzyapko O, Melkov G A, Serga A A, Hillebrands B and Slavin A N 2006 *Nature* **443** 430
[56] Demidov V E, Dzyapko O, Buchmeier M, Stockhoff T, Schmitz G, Melkov G A and Demokritov S O 2008 *Phys. Rev. Lett.* **101** 257201
[57] Dzyapko O, Demidov V E, Demokritov S O, Melkov G A and Safonov V L 2008 *Appl. Phys. Lett.* **92** 162510
[58] Chumak A V, Melkov G A, Demidov V E, Dzyapko O, Safonov V L and Demokritov S O 2009 *Phys. Rev. Lett.* **102** 187205
[59] Rezende S M 2010 *J. Phys.: Condens. Matter* **22** 164211

Appendix

Some mathematical topics

Two of the topics that were mentioned briefly in sections 1.4 of chapter 1 are now described here in more detail. These are, first, the TDM method as used in discrete-lattice calculations when there are only short-range interactions and, second, the linear-response method along with its formal connection to the quantum-mechanical Green-function techniques of many-body theory.

A.1 The tridiagonal matrix method

Often, in solving dynamical problems involving discrete-lattice systems and including cases with one or more surfaces or interfaces, we may encounter TDMs if the interactions between lattice points are sufficiently short range. In general TDMs are square matrix arrays, either finite or infinite in dimension, in which the only nonzero elements are along the leading diagonal and the two adjacent diagonals above and below. These matrices have some convenient analytic properties, particularly if there is some pattern to the values of the nonzero elements.

Some early applications of TDMs for the dynamics of surfaces are to be found in work by Wallis [1] for phonons and De Wames and Wolfram [2] for magnons or SWs, while some formal properties were elucidated in [3]. Let us consider, for example, the infinite set of finite-difference equations corresponding to equation (2.3) for a monatomic chain, which may easily be recast in a matrix form as $\mathbf{A} \mathbf{u} = 0$, where $\mathbf{u}$ is the column matrix with elements u_1, u_2, u_3 and so on. After dividing throughout by the appropriate factors, the matrix $\mathbf{A}$ is

$$\mathbf{A} = \begin{pmatrix} d_1 & -1 & 0 & 0 & \cdots \\ -1 & d & -1 & 0 & \cdots \\ 0 & -1 & d & -1 & \cdots \\ 0 & 0 & -1 & d & \cdots \\ \vdots & \vdots & \vdots & \vdots & \ddots \end{pmatrix}, \tag{A.1}$$

doi:10.1088/978-0-7503-1054-3ch8

where $d = (2C - m\omega^2)/C$ and there is just the one perturbed matrix element $d_1 = d - 1$ (modified due to the surface), which is at the top left corner of the array.

In the above example we have a simple expression for d_1, but for a more general model of a surface, involving possibly several modified parameters, we could write

$$\mathbf{A} \mathbf{u} = (\mathbf{A}_0 + \mathbf{D})\mathbf{u} = 0 \qquad (A.2)$$

with $\mathbf{A}_0$ denoting the unperturbed dynamic matrix

$$\mathbf{A}_0 = \begin{pmatrix} d & -1 & 0 & 0 & \cdots \\ -1 & d & -1 & 0 & \cdots \\ 0 & -1 & d & -1 & \cdots \\ 0 & 0 & -1 & d & \cdots \\ \vdots & \vdots & \vdots & \vdots & \ddots \end{pmatrix}, \qquad (A.3)$$

and with $\mathbf{D}$ representing the perturbation matrix due to the presence of the surfaces and/or interfaces. It can be shown that the inverse matrix $\mathbf{A}_0^{-1}$ is given explicitly by

$$\left(\mathbf{A}_0^{-1}\right)_{i,j} = \frac{x^{i+j} - x^{|i-j|}}{x - x^{-1}}, \qquad (A.4)$$

where x is a complex variable defined by

$$x + x^{-1} = d, \quad (|x| \leqslant 1). \qquad (A.5)$$

The above result can be verified, for example, by showing explicitly that it satisfies $\mathbf{A}_0^{-1}\mathbf{A}_0 = \mathbf{I}$, where $\mathbf{I}$ is the unit matrix. Now equation (A.2) can be manipulated into the form

$$\mathbf{A}_0\left(\mathbf{I} + \mathbf{A}_0^{-1}\mathbf{D}\right)\mathbf{u} = 0. \qquad (A.6)$$

If there is no perturbation term ($\mathbf{D} = 0$), the determinantal condition for a non-trivial solution is simply $\det \mathbf{A}_0 = 0$. We see, however, that when $\mathbf{D}$ is nonzero there may also be solutions arising from

$$\det\left(\mathbf{I} + \mathbf{A}_0^{-1}\mathbf{D}\right) = 0. \qquad (A.7)$$

In fact, the solutions of equation (A.7) with $|x| < 1$ will correspond to the localized surface and/or interface excitations of the system. In contrast, the solutions of $\det \mathbf{A}_0 = 0$ will generally be associated with the bulk modes and correspond to complex values of x on the unit circle (with $|x| = 1$).

Equation (A.7) provides a very useful result for determining the existence conditions and dispersion relations for the localized modes at surfaces and/or interfaces. This because $\mathbf{D}$ has typically only a few nonzero matrix elements and so the dimension of the determinant becomes small. For example, it will sometimes be the case that there is just one matrix element of $\mathbf{D}$, say $D_{1,1}$ as in the example mentioned above, corresponding to the surface of a semi-infinite medium. It is then easily shown that equation (A.7) reduces to $(1 + xD_{1,1}) = 0$. Hence there will be one surface mode for x provided the localization condition, which becomes

$$|D_{1,1}| > 1, \qquad (A.8)$$

can be satisfied. If this is so, the dispersion relation is found by substituting $x = -1/D_{1,1}$ into equation (A.5) and using the definition of d. In the case cited earlier for the longitudinal vibrational modes of a terminated semi-infinite monatomic chain (equation (A.1) and subsection 2.1.1) we found $D_{1,1} = 1$ and so there is no surface mode in this case, as was stated in subsection 2.1.1.

On the other hand, for the vibrational modes in a semi-infinite diatomic chain (as discussed in subsection 2.1.2) we have the two sets of linear coupled equations given in equation (2.8) for the odd- and even-numbered masses, together with equation (2.10) for the 'surface' $n = 0$ mass. The TDM method may be straightforwardly employed in such cases by using the odd equations to eliminate the odd displacement terms in the even set of equations, or vice versa. For example, after some simple algebraic manipulation we find for the semi-infinite chain that

$$[w_1(w_2 - 1) - 1]u_0 = u_2 \quad (n = 0),$$
$$(w_1 w_2 - 2)u_{2n} = u_{2n-2} + u_{2n+2} \quad (n \geqslant 1), \tag{A.9}$$

where we denote $w_1 = 2 - (m_1\omega^2/C)$ and $w_2 = 2 - (m_2\omega^2/C)$. The equations can be re-expressed in terms of a tridiagonal dynamic matrix having the same form as equation (A.3) with the redefinitions

$$d = w_1 w_2 - 2, \qquad D_{1,1} \equiv d_1 - d = 1 - w_1. \tag{A.10}$$

We may then follow the formal steps described in equations (A.4)–(A.8) to obtain results for the phonon frequencies. The solutions with $|x| = 1$ yield the 'bulk' modes of the system: specifically, putting $x = \exp(iqa)$ where q is the wavenumber yields $w_1 w_2 - 2 = 2\cos(qa)$, which is found to be equivalent to the result quoted in equation (2.11). The surface-mode solution comes from $x = -1/D_{1,1}$, provided equation (A.8) is satisfied, and leads to the result for ω_S quoted in equation (2.11).

Other physical examples where the semi-infinite TDM formalism can usefully be applied arise in the context of the surface SWs in exchange-coupled ferromagnetic and antiferromagnetic systems and the electronic edge modes in graphene, which were described in chapters 3 and 4, respectively. The calculation for a Heisenberg ferromagnet was given in subsection 3.2.1 and the coupled equations in terms of a layer index n appear in equation (3.9). For a thick film (taking $N \to \infty$) the dynamic equation can be rewritten in the same matrix form as in equations (A.1)–(A.3), but with $\mathbf{u}$ now denoting the column matrix with elements given by the spin amplitudes $s_1, s_2, s_3, \ldots$ and with the previous definitions of d and $D_{1,1} \equiv (d_1 - d)$ replaced by

$$d = \frac{(g\mu_B B_0 - \omega)}{SJ} + 2SJ\left[3 - 2\gamma\left(\mathbf{q}_\parallel\right)\right],$$
$$D_{1,1} = -\left(1 + \frac{4(J - J_s)}{J}\left[1 - \gamma\left(\mathbf{q}_\parallel\right)\right]\right), \tag{A.11}$$

in the notation of chapter 2. The definition of the complex x in equation (A.5) still applies and it may easily be verified that the solutions for ω for which $|x| = 1$ (putting $x = \exp(iq_z a)$) yields the previous bulk SW solution in equation (3.12), as expected. Also, by analogy with the discussion for surface phonons earlier in this section, the TDM method indicates that a surface SW exists provided the condition in equation (A.8) is satisfied (ensuring $|x| < 1$) and then we have a solution for $x = -1/D_{1,1}$. It is a straightforward algebraic exercise to verify that the results quoted for the surface SW in subsection 3.2.1 are recovered, noting that the surface parameter Δ in equation (3.14) can be equated to $-1/D_{1,1}$. The case of applying the TDM method to SW modes in bct Heisenberg antiferromagnets (see subsection 3.2.2) can be treated using a similar approach to that for the vibrational modes (phonons) in a diatomic chain. For graphene with a zigzag edge we already made use of the TDM method in section 4.2, where the corresponding expressions for d and $D_{1,1}$ are quoted.

The formalism is straightforwardly extendible to finite N, as needed for finite-length chains and finite-thickness films. We again use $\mathbf{A} = (\mathbf{A}_0 + \mathbf{D})$, where $\mathbf{A}_0$ is now the $N \times N$ matrix given by

$$
\mathbf{A}_0 = \begin{pmatrix}
d & -1 & 0 & \cdots & 0 & 0 \\
-1 & d & -1 & \cdots & 0 & 0 \\
0 & -1 & d & \cdots & 0 & 0 \\
\vdots & \vdots & \vdots & \ddots & \vdots & \vdots \\
0 & 0 & 0 & \cdots & d & -1 \\
0 & 0 & 0 & \cdots & -1 & d
\end{pmatrix}. \tag{A.12}
$$

Its inverse in this case is given by (see [4])

$$
\left(\mathbf{A}_0^{-1}\right)_{i,j} = \frac{x^{i+j} - x^{|i-j|} + x^{2N+2-(i+j)} - x^{2N+2-|i-j|}}{\left(1 - x^{2N+2}\right)\left(x - x^{-1}\right)}. \tag{A.13}
$$

It follows from the definition of x in equation (A.5) that terms of order x^{2N} vanish as $N \to \infty$, making the above result reduce to equation (A.13).

A.2 Linear-response and Green functions

Here we provide some additional steps needed to obtain the results quoted in subsection 1.4.2. There we considered a perturbation involving an operator B coupled linearly to a time-dependent scalar field $f(t)$ as in equation (1.37) and we wanted to find the corresponding time-dependent linear response produced in a system variable described by any other operator A.

To achieve this we can use an important result from statistical mechanics that the mean value of any operator A is worked out from $\mathrm{tr}(\rho A)$ where 'tr' denotes the trace of the operators and ρ is the density matrix of the system [5, 6]. The trace of an operator is just the sum of all the diagonal elements if a matrix representation is used.

Before the perturbation is switched on at time $t = -\infty$ the density matrix of the unperturbed system in equilibrium is given by

$$
\rho_0 = \frac{\exp(-H_0/k_B T)}{\mathrm{tr}[\exp(-H_0/k_B T)]}, \tag{A.14}
$$

for a canonical ensemble [5, 6] with T denoting the temperature and k_B the Boltzmann constant. At a later time t the density matrix of the system will be changed to $\rho = \rho_0 + \rho_1$, where ρ_1 can be deduced using the equation of motion for the density matrix as

$$\rho_1 = i \int_{-\infty}^{t} \exp\{iH_0(t' - t)\}[B, \rho_0]\exp\{-iH_0(t' - t)\}dt', \qquad (A.15)$$

and the right-hand side has been linearized in operator B. The required result for the linear response $\bar{A}(t)$ can eventually be manipulated into the form (see [8])

$$\bar{A}(t) = i \int_{-\infty}^{t} \langle[A(t), B(t')]\rangle f(t')dt', \qquad (A.16)$$

where $\langle\cdots\rangle = tr(\rho_0\cdots)$ indicates a thermal average for the unperturbed system, while $A(t)$ is a shorthand for $\exp(iH_0t)A\exp(-iH_0t)$ and similarly for $B(t')$.

With equation (A.16) as the main linear-response result we are now in a position to explain the connection with Green functions as follows. For any two operators A and B we may formally define a Green function $\langle\langle A(t); B(t')\rangle\rangle$ in a time representation as

$$\langle\langle A(t); B(t')\rangle\rangle = i\theta(t - t')\langle[A(t), B(t')]\rangle, \qquad (A.17)$$

where, specifically, apart from the overall sign, this is equivalent to the two-time commutator Green function introduced by Zubarev [7]. We employ the notation

$$\theta(t - t') = \begin{cases} 1 & \text{for } t > t', \\ 0 & \text{for } t < t', \end{cases} \qquad (A.18)$$

for the unit step function. Its appearance as a factor in equation (A.17) is in accordance with causality (in the sense that the stimulus applied through operator B at time t' produces a response in operator A at a later time t). Equation (A.16) can now be re-expressed as

$$\bar{A}(t) = i \int_{-\infty}^{\infty} \langle\langle A(t); B(t')\rangle\rangle f(t')dt'. \qquad (A.19)$$

This result, however, is more useful when transformed from the time to the frequency representation. It can be shown that the Green function depends on the time labels only through the difference $(t - t')$ and so we may take

$$\langle\langle A(t); B(t')\rangle\rangle = \int_{-\infty}^{\infty} \langle\langle A; B\rangle\rangle_\omega \exp[-i\omega(t - t')]d\omega, \qquad (A.20)$$

where the quantities $\langle\langle A; B\rangle\rangle_\omega$ are the frequency Fourier components of the Green function. Finally, if a similar definition is made for $\bar{A}_\omega$ and $F(\omega)$ as the Fourier components of $\bar{A}(t)$ and $f(t')$, respectively, we find that equation (A.19) simplifies to

$$\bar{A}_\omega = \langle\langle A; B\rangle\rangle_\omega F(\omega) \qquad (A.21)$$

in the frequency representation. This last result clearly shows the role of Green functions as linear-response functions, in the sense that they are just the same as the proportionality factor between an applied stimulus to the system, represented by $F(\omega)$, and the response that it produces, represented by $\bar{A}_\omega$.

Two other results for Green functions (see, e.g., [8]) are worth quoting here. First they satisfy an equation of motion of the form

$$\omega\langle\langle A; B\rangle\rangle_\omega = (1/2\pi)\langle[A, B]\rangle + \langle\langle[A, H]; B\rangle\rangle_\omega \qquad (A.22)$$

and, secondly, the correlations in the dynamical system are related to the imaginary part of a corresponding Green function by

$$\langle BA\rangle_\omega = \frac{1}{\pi[1 - \exp(-\omega/k_BT)]}\text{Im}\,\langle\langle A; B\rangle\rangle_\omega. \qquad (A.23)$$

The first of these results is sometimes useful in calculating the Green function, as an alternative to deducing it from equation (A.21). The second result is often referred to as the *fluctuation-dissipation theorem*. It can be employed to calculate the spectral intensities of the excitations in a system and also scattering cross sections.

References

[1] Wallis R F 1957 *Phys. Rev.* **105** 540
[2] de Wames R E and Wolfram T 1969 *Phys. Rev.* **185** 720
[3] Rutherford D E 1947 *Proc. R. Soc. Edinburgh* A **62** 229
[4] Cottam M G and Kontos D E 1980 *J. Phys. C: Solid State Phys.* **13** 2945
[5] Landau L D and Lifshitz E M 1080 *Statistical Physics* (Oxford: Pergamon)
[6] Reichl L E 2009 *A Modern Course in Statistical Physics* 3rd edn (Weinheim: Wiley)
[7] Zubarev D N 1960 *Sov. Phys.—Usp* **3** 320
[8] Cottam M G and Tilley D R 2005 *Introduction to Surface and Superlattice Excitations* 2nd edn (Bristol: IOP)

Lightning Source UK Ltd.
Milton Keynes UK
UKHW05n0614180618
324301UK00003B/70/P